EARTH 2020

Earth 2020

*An Insider's Guide to
a Rapidly Changing Planet*

Philippe Tortell

OpenBook
Publishers

https://www.openbookpublishers.com

ISBN Paperback: 978-1-78374-845-7

ISBN Hardback: 978-1-78374-846-4

ISBN Digital (PDF): 978-1-78374-847-1

ISBN Digital ebook (epub): 978-1-78374-848-8

ISBN Digital ebook (mobi): 978-1-78374-849-5

ISBN Digital (XML): 978-1-78374-850-1

DOI: 10.11647/OBP.0193

Cover image: *Earthrise* (24 December 1968). Photo taken by Apollo 8 crewmember Bill Anders, Wikimedia, https://commons.wikimedia.org/wiki/File:NASA_Earthrise_AS08-14-2383_Apollo_8_1968-12-24.jpg

Cover design: Anna Gatti

Contents

Epigraph

Whoa, ah, mercy mercy me
Oh things ain't what they used to be, no no
Where did all the blue skies go?
Poison is the wind that blows from the north and south and east

Whoa mercy, mercy me,
Oh things ain't what they used to be, no no
Oil wasted on the oceans and upon our seas, fish full of mercury

Ah, oh mercy, mercy me
Ah things ain't what they used to be, no no
Radiation underground and in the sky
Animals and birds who live nearby are dying

Oh mercy, mercy me
Oh things ain't what they used to be
What about this overcrowded land
How much more abuse from man can she stand?

Marvin Gaye, 'Mercy Mercy Me (The Ecology)',
What's Going On (Motown Records, 1971).

Acknowledgements

I am grateful, first and foremost, to all of the authors who contributed their words and insights to this volume. They worked under tight deadlines, juggling many other commitments, and were exceedingly patient with all of my many editorial suggestions. Through their eyes, I have come to a much deeper understanding of Earth's environmental history over the past half century. I wish also to thank the group at Open Book Publishers, whose outstanding (and very rapid) work made this book a reality. Alessandra Tosi was enthusiastic about this project from the start, and marshaled the team to keep things moving forward throughout. Both Alessandra and Adèle Kreager provided excellent editorial suggestions, while Luca Baffa developed a beautiful design and layout, and Laura Rodriguez helped spread the word about this project far and wide. During the early phases of this project, Margot Young, Mark Turin and Jesse Finkelstein provided much needed advice and inspiration. Finally, I wish to thank my family for support and patience as I set aside weekends, evenings and early mornings to find time to work on this project amidst a sea of other responsibilities. I hope they agree that the result was well worth the effort.

Introduction

—

Philippe Tortell

On 22 April, 1970, millions of people took to the streets in cities and towns across the United States, giving voice to an emerging consciousness of humanity's impact on planet Earth. This first Earth Day was the brainchild of US Senator Gaylord Nelson, and was organized by a grassroots movement coordinated by Denis Hayes, a twenty-five-year-old Harvard student. The Earth Day events included demonstrations, teach-ins and community clean-ups ('Trash Wednesday') in over 2,000 communities across the country. Protesters shut down Fifth Avenue in New York City, while students in Boston staged a 'die-in' at Logan airport, lying in coffins to raise awareness about the dangers of airplane-related pollution. Demonstrators in Chicago called for an end to the internal combustion engine. The protesters were mostly white, middle-class and overwhelmingly young, but their message also reached some in the older generation. Walter Cronkite, by then widely seen as the most trusted man in America, hosted a half-hour Earth Day special on the CBS Evening News. He had become increasingly concerned about 'the fouled skies, the filthy waters and the littered earth', as he put it, and he concluded the news special with a call for the public to heed 'the unanimous voice of the scientists warning that half-way measures and business as usual cannot possibly pull us back from the edge of the precipice'.

Today, half a century later, Cronkite's words are eerily familiar. Since the first Earth Day, we have, no doubt, made significant progress in addressing a range of acute environmental problems. Yet, other more pernicious threats have emerged, from climate

https://doi.org/10.11647/OBP.0193.01

change to global biodiversity loss; the warnings seem louder, and the edge of the precipice ever closer, as growing evidence demonstrates planetary-scale human perturbations of the Earth System. As we look back to the first Earth Day fifty years ago, understanding the environmental trajectory of planet Earth, and the societal evolution of its most dominant species — humans — provides us with important lessons from the past, and, hopefully, insights for the future.

Such lessons and insights are gathered together in the present collection of essays, which mark the fiftieth anniversary of Earth Day in 2020. The idea for such a collection first came to me in January 2019, just a few months after I had resigned as Director of the Institute for Advanced Studies at the University of British Columbia (UBC). It was a job I had held for the better part of three years, during which I had worked with scholars from across the university and around the world, fostering inter-disciplinary research on a wide range of topics. As part of this work, I had co-edited two collections of essays, *Reflections of Canada* and *Memory* marking, respectively, the 150th anniversary of Canadian Confederation in 2017, and the 100th anniversary of the end of WWI in 2018. These projects brought leading scholars together to share their insights on those historical milestones, in lively and accessible prose aimed at a broad audience. Anniversaries, I had learned, provided a valuable opportunity to focus public attention (if only for a short while) on topics of significant importance.

As I resumed my duties as a full-time professor of oceanography at UBC, I found myself with a sense of restlessness, and a desire to think beyond the bounds of a single academic field. During this time, I stumbled across the Earth Day network (https://www. earthday.org/earth-day-2020/). I was well aware of Earth Day, and had even (at least some years), participated in the event in some small symbolic way. I recall more than once sitting around a candle lit dinner table with friends as we turned off the lights for the prescribed hour. Maybe some of my neighbors did the same; maybe not. It was a ritual nod to our environmental consciousness, but we had little understanding of the origins and historical significance of this event, which had begun two years before I was born.

And yet, I and many others had become increasingly concerned about a range of growing environmental problems. During the mid-1990s, I was a PhD student in the US when the UN's Intergovernmental Panel on Climate Change (IPCC) released its Second Assessment Report, which asserted, for the first time, strong evidence for a discernable human impact on global climate.[1] This report galvanized many around the world, on all sides of the debate. On the one hand, scientific advances and rapidly increasing computer power were providing fundamental new insights into global climate dynamics, and vastly improved predictive capabilities that enabled us to glimpse into the possible future of our planet. On the other hand, powerful forces were marshaling against science, backed by well-funded industry groups with vested interests in the status quo, who sought to exploit legitimate scientific uncertainty to argue against meaningful climate change mitigation. During my third year of graduate school, in 1997, the nations of the world developed a joint framework to limit global greenhouse emissions under the Kyoto Protocol.[2] This landmark agreement continued the legacy of the Earth Summit in Rio just five years earlier, and represented the first steps towards tackling climate change. Unfortunately, the aspirations of Kyoto (and Rio, for that matter) unraveled quickly; about two months before I obtained my PhD, in the spring of 2001, US President George W. Bush announced that the US would not implement the 1997 Kyoto Protocol. Many other nations soon followed suit, signaling the death-knell of the agreement. Over the short span of my graduate education, I had witnessed a radical shift in global environmental politics.

The following year, in 2002, I began a research and teaching career at UBC. Among other things, my work focused on understanding the potential effects of rising ocean CO_2 levels (and decreasing pH) on the productivity of phytoplankton (microscopic plants at the base of the marine food chain). Over the next fifteen years, I conducted ship-based studies of the global ocean, from the tropics to the poles, including multiple research expeditions to the Arctic and Antarctic regions. These experiences left a strong impression. I witnessed, first hand, the human footprint on marine ecosystems; from rapidly retreating glaciers and sea ice, to warming and acidifying ocean surface waters, and plastic debris floating thousands of kilometers from the shore. At the same time, interactions with my colleagues across a range of disciplines at UBC deepened my understanding of the changes that were rapidly unfolding across other parts of the Earth System, including agricultural lands and forests, wetlands, lakes and rivers. And I also knew that society was evolving, with increasing awareness of growing environmental challenges, and shifting narratives around sustainable resource use and meaningful engagement with Indigenous peoples. In British Columbia, where the economy is strongly dependent on extractive resource industries (mining, forestry and, increasingly, natural gas), there was much debate over how to balance economic development with environmental stewardship and reconciliation with First Nations, on whose traditional lands much of the provinces resource base is located.

Once I had decided to assemble an Earth Day-themed book, I began by looking back to the late 1960s and early 1970s, trying to better understand the historical context that led to the events that unfolded on April 22, 1970. Highlights included reading the original series of three *New Yorker* articles where Rachel Carson first published excerpts of her classic environmental book, *Silent Spring*,[3] and watching the thirty-minute CBS News Earth Day special with Walter Cronkite. As I was learning more about the history of Earth Day, a visiting colleague also pointed me to the 1971 book, *Patient Earth*, by John Harte and Robert Socolow.[4] The book provides a remarkably prescient and in-depth examination of early thinking around the nascent field of environmental science, with chapters on a range of topics, including human population growth, resource scarcity, nuclear power, land-use conflicts and steady state economics. I wondered what such a book would look like, had it been written fifty years later, in 2020. And so, taking inspiration from *Patient Earth*, and from my own experiences at the Institute for Advanced Studies, I sought to better understand, from a wide range of perspectives, how Earth's biophysical systems had been impacted by anthropogenic activities over the past half-century, and how society had evolved to mitigate (or perhaps exacerbate) the human environmental footprint. From this starting point, I quickly arrived at a number of topics, and sought world-leading experts from many disciplines who could address these with authority and eloquence. The response from prospective authors was overwhelmingly positive; almost everyone I contacted agreed to contribute to the project. The result is the collection of ideas and words you now hold in your hands (or read on a digital screen).

From the outset, the thematic structure of the book was clear enough. I asked all authors to reflect, from their own vantage point, on how Earth and its human inhabitants had changed over the past fifty years, and what the future might look like another fifty or more years down the road. The contributions fell naturally into several groups. Some authors examined biophysical components of the Earth System, from the atmosphere (Jon Abbatt), oceans (David M. Karl) and fresh waters (Janet G. Hering), to glaciers (Julian Dowdeswell), land (Navin Ramankutty and Hannah Wittman) and forests (Sally N. Aitken), while others examined impacts on organisms and ecosystems, presenting case studies of declining marine fisheries (U. Rashid Sumaila and Daniel Pauly) and dwindling global biodiversity, writ large (Jeffrey R. Smith and Gretchen C. Daily). Still other essays addressed the pernicious problem of long-lived wastes in the form of plastics (Roland Geyer), toxic chemicals (Elsie Sunderland and Charlotte C. Wagner) and space junk (Alice Gorman) that are a defining feature of the Anthropocene — a new geological era dominated by human influence on planet Earth. Other authors looked at the shifts in political (Elizabeth May), legal (Rosemary Lyster) and economic (Don Fullerton) paradigms that have occurred since 1970, as well as the evolving media landscape in which all of these changes have unfolded (Candis Callison) and the role of science and technology in shaping societal actions and discourse (Sheila Jasanoff).

At a global scale, there is no doubt that increasing human consumption of fossil fuels has driven a large-scale perturbation of the global climate system. One essay on carbon (David Archer) explores this anthropogenic footprint in the context of deep geologic time, while another (Elizabeth J. Wilson and Elias Grove Nielsen) examines the underlying global energy trends driving historic and potential future CO_2 emissions. A deeper understanding of the impacts of rising CO_2 on the climate has only emerged in recent decades, as shown by an essay on the historical development of computer climate models (Tapio Schneider), and essays discussing recent and possible future trends in global sea level (Robert E. Kopp) and extreme weather (Neville Nicholls). And as these impacts become ever clearer, there is increasing discussion of potential geoengineering to limit the worst potential consequences, as discussed in one essay (Douglas G. MacMartin and Katharine L. Ricke). These technological approaches represent a case of fighting fire with fire, but perhaps there are other ways to imagine the problem and its potential solutions. In this respect, long-held wisdom of Indigenous knowledge systems (Deborah McGregor) has much to teach us. At the same time, other 'world-views' can be brought to bear, using audio and visual media to re-frame our world through the lens of the creative arts (Edward Burtynsky).

Despite the diversity of ideas and topics presented in this collection, there are some gaps. Global population growth is a prime example. In the late 1960s and early 1970s, rapid improvements in public health led to sharp increases in longevity that were not matched by declining birth rates, leading some, like Stanford biologist Paul R. Ehrlich, to raise awareness of an impending 'population bomb', as he titled his 1968 book.[5] Today, global population now exceeds seven billion (roughly double what it was in 1970) and more than half of all the planet's inhabitants now live in cities. But this overall population growth has been accompanied by a large demographic transition, with populations falling in some countries. In turn, girls and women have increasingly widespread access to education and reproductive controls to limit unwanted pregnancies. Even if the most direct projections in Ehrlich's book do not come to pass, there can be no doubt that the growing human population has put an increasing burden on Earth's planetary systems. Other topics not addressed in *Earth 2020* include the rise of global pandemics and antibiotic resistance, both of which could have significant environmental impacts on human societies. Clearly, these topics warrant further attention and research. Similarly, environmental justice, which is touched on by several authors in this collection (May, Lyster and McGregor), deserves more in-depth treatment, as climate change and ecological degradation disproportionately affect some of the world's most vulnerable populations.

The solutions to our environmental problems cannot be siloed into distinct domains of expertise, and this is reflected in the integrated approach of many of the authors in this collection, who explore ideas that cross traditional boundaries, as well as in the structure of this volume. Instead of being organized thematically, grouped by discipline and subject matter, essays on different topics are scattered throughout the book, like stepping stones across a stream of ideas, with many possible ways to cross. But the banks of the stream — the beginning and ending crossing points — were clear from the beginning. It seemed only natural to start with an explicit retrospective (John Harte and Robert Socolow), going back to *Patient Earth* to re-examine long-standing environmental questions with the benefit of hindsight. The other side of the stream represents our unknown future. What will the planet look like in 2070, and how will our current understanding of Earth's trajectory map onto the reality that unfolds over the next half-century? Few of the authors in *Earth 2020* will be able to answer this question; many are at, or approaching, the end of their careers, and few will even be around to see 2070. Nor will they be the ones most burdened by the environmental consequences of our collective actions over the past fifty years. For this reason, the last word must be given to our newest generation of leaders (Zoe Craig-Sparrow and Grace Nosek), those who have stepped up to demand systemic change, and who will drive the way, with our support and encouragement, to a better future.

As we look to the uncertain future ahead, it is clear that our path forward will not resemble the road we have traveled to get here. As the essays in this book demonstrate, planet Earth has changed in profound ways, and these changes will be with us for generations to come. In the face of this transformation, we must not be paralyzed by fear and anxiety. Rather, we must harness new tools and understanding, working collectively to develop innovative approaches to address many of our most challenging environmental and social problems. In that respect, free and open exchange of ideas and information is critical; we must be able to learn from each other, drawing inspiration from past successes, while avoiding previous mistakes. It seemed only natural, therefore, to use an open access publishing model for *Earth 2020*, making it freely available to anyone in the world. But wide distribution is not enough. We must also explore other multimodal approaches to engage broad audiences who feel increasingly overwhelmed in the age of information overload, where ideas compete for relevance in a crowded digital landscape. To this end, two examples of multimodality are offered as part of this volume, in the section directly following this introduction. These take the form of musical compositions drawn from a range of Earth System data; sonic representations of our rapidly evolving planet.

For much of the past year, as I have worked on this book, my own outlook on planet Earth has fundamentally shifted. For one thing, I have come to a much deeper understanding of the historical and political context that has driven humanity's impact on the planet. Through the words and ideas of the book's authors, the events that have unfolded around me over the past five decades have come into sharper focus as part of a larger emergent narrative. And what stands out most, perhaps, is the notion of possibility. It is true that things look grim, but they also did in 1970. Our history has shown that we have the capability to address daunting global challenges if we have the will and the fortitude. In the words of the young climate activist, Greta Thunberg, delivered to the US Congress, in September, 2019: 'You must take action. You must do the impossible. Because giving up can never ever be an option'. It is my great hope that you, the reader, will find both knowledge and inspiration in this book, and that it will mobilize you to take action in pushing society towards a more just and sustainable future.

Endnotes

1. Available at https://www.ipcc.ch/site/assets/uploads/2018/06/2nd-assessment-en.pdf

2. Available at https://unfccc.int/resource/docs/convkp/kpeng.pdf

3. R. Carson, *Silent Spring*, New York: Houghton Mifflin, 1962.

4. J. Harte and R. Socolow, *Patient Earth*, New York: Holt, Rinehart and Winston, 1971.

5. P. R. Ehrlich, *The Population Bomb*, New York: Ballantine Books, 1968.

Earth Sounds

—

Philippe Tortell, Chris Chafe, Jonathan Girard and Greg Niemeyer

*I**ce Core Walk* is a musical representation of environmental climate data taken from the 3 km-long Vostok ice core in East Antarctica. The audio clip below represents a snapshot of atmospheric temperature and CO_2 data, from 850 AD to 2016, translated into musical form. This clip is taken from the last five minutes of a half-hour-long audio tour, which allows listeners to experience 800,000 years of climate history as they walk the full 3 km-length of the Vostok ice core. The most recent temperature data are obtained from tree ring measurements, sediments and other sources, while the CO_2 data are from a combination of the National Center for Atmospheric Research (NCAR) Community Climate System Model simulations and direct observations. The sounds are synthesized from a physical model of a plucked nylon string guitar — indicating temperature — and a vocal-like synthesis — indicating CO_2 levels. This composition articulates the pace of climate change sonically rather than visually, offering listeners a stark audio-perspective on the impacts of humans on the climate system over the past century. *Ice Core Walk* is a collaboration between scientists and artists from the University of British Columbia (UBC), Stanford University and the University of California, Berkeley. The project was initiated by Philippe Tortell,[1] Chris Chafe[2] and Greg Niemeyer[3] and was supported by the Peter Wall Institute for Advanced Studies, UBC. More information about *Ice Core Walk* can be found at http://icecorewalk.org/, along with the full half-hour-long audio tour.

          https://doi.org/10.11647/OBP.0193.02

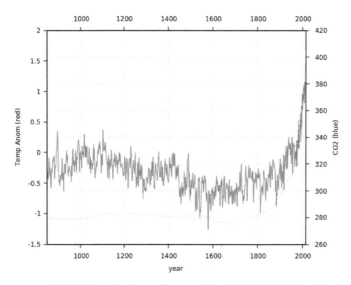

Temperature and CO2 data in the ending section (850 AD–2016) of Ice Core Walk.
CC BY-NC-SA 3.0 US. See web site for data sources,
https://purl.stanford.edu/mg458wc3389

Ice Core Walk

© 2016 Philippe Tortell, Chris Chafe and Greg Niemeyer, CC BY 4.0

https://doi.org/10.11647/OBP.0193.29

E *arth Symphony* is a musical representation of our planet's trajectory over the past fifty years, drawn from a range of Earth System data sets — from atmospheric CO_2 concentrations and global fisheries catches, to deforestation and the size of the Antarctic ozone hole. These data sets have been translated into a musical score, using a process of sonification that seeks to express and more deeply understand the complex biophysical changes unfolding across the Earth System. The piece is an interdisciplinary collaboration between students and scholars: Philippe Tortell compiled these data from public sources with the help of Environmental Sciences undergraduate students at UBC, and Chris Chafe 'sonified' the data into a musical score by creating a process in which music is performed directly by the data. His choices as composer included the speed at which the data are heard, the instruments that play the sounds, and the influence of the data on musical dimensions like pitch and loudness. In recognition of the fiftieth anniversary of Earth Day in 2020, *Earth Symphony* will be performed by the UBC Orchestra, conducted by Jonathan Girard.[4] A video recording of this performance will be available at planetearth2020.org

Earth Symphony

https://doi.org/10.11647/OBP.0193.30

Endnotes

1. Professor of Oceanography at UBC, and editor of this volume.

2. Director of Stanford University's Center for Computer Research in Music and Acoustics.

3   Director of Orchestras at the UBC School of Music.

4   Professor of New Media in Art Practice at the Univeristy of California, Berkeley.

Impatient Earth

—

John Harte and Robert Socolow

Fifty years ago, the two of us wrote an introductory textbook, *Patient Earth*, about a range of environmental problems that were coming into focus as we entered the final decades of the twentieth century.[1] Our book told its story partially through ten contemporary site-specific case studies, which were chosen based on their likely staying power — would they be relevant in fifty years? All of them are.

Recently, we discussed the need for a new, updated *Patient Earth*, in which fifty intervening years of developments in environmental science and policy would be presented. Soon thereafter, we met Philippe Tortell and discovered that he was gearing up to write just such a book. *Earth 2020*, as he described it, would cover a comprehensive set of topics, with chapters authored by global experts in each field. We were thrilled to be asked to contribute some perspectives to this timely book, which we expect to be relevant still, half a century from now.

Comparing and contrasting our book with this present volume, *Earth 2020*, can teach all of us a lot about how the world has changed over the past half-century, and what the future may yet hold. For one thing, *Patient Earth* was the product of a white, male, upper-class world, with only two female authors, and an antediluvian treatment of pronouns. For another thing, *Patient Earth*, unlike *Earth 2020*, could not have looked back fifty years. In 1970, environmentalism had much less of a past than it does today. At that time, it was

https://doi.org/10.11647/OBP.0193.03

a frontier; now, it is mainstream. We did have an essay by Paul Sears that looked back nearly fifty years to the Dust Bowl calamity of the 1930s, and considered 'the inseparable tie between the good earth and human destiny'.[2] We paired that essay with another, by Jeremy Sabloff, that looked even further back, to the collapse of the Maya civilization.[3] The word 'sustainability' hardly existed in 1970, but these two essays did call attention to risks to the continuity of civilization.

In our introduction to those two 'Lessons from the Past', we noted that the Dust Bowl tragedies resulted from farmers, ranchers and land developers ignoring the warnings of soil scientists and agronomists. The Maya, we suggested, did not see the consequences of their population growth under limited land resources, and lacked the knowledge to make the metal tools that might have extended their farmland. We wrote: 'Every society has its blind spots and from a distance one's reactions to them are instinctively charitable. But to the deaf spots in a society, how should one respond?'[4]

Let us turn that judgmental spotlight upon ourselves, and assess our choices of topics in *Patient Earth*. Which warnings did we hear, which could we have heard if we had paid attention, and which did we not hear because they did not yet exist? Such analysis can provide insight, more generally, into how society can learn to open its ears.

In 1970, environmentalism was deeply intertwined with three other contemporary concerns: wilderness and the non-human environment, militarism and population. We were determined to address all three. Notably, they are scarcely present in the collection of topics addressed in *Earth 2020*.

To emphasize wilderness and the non-human environment, we recruited an essay by Albert Hill and Michael McCloskey about how the High Sierras in California were about to be invaded by a ski resort,[5] and another by Kent Shifferd about how the remote woods of northern Wisconsin were threatened by an immense transmitter for submarine communications.[6] We also wrote our own essay on the menace to the Florida Everglades presented by a proposed international jetport west of Miami.[7] Activists battled all three, and none were built. Today, environmental organizations present the need to protect the environment in largely instrumental terms, stressing the direct benefits to humans (clean air and water, and carbon storage, for example). We straddled this breech ourselves. In our essay on the Everglades, we highlighted the negative human impacts resulting from the degradation of nature and noted how 'the well-being of man (*sic*) and the park, in quite direct and material ways, are critically linked',[8] a notion now referred to as 'ecosystem services.' But we could not have guessed then that fifty years later, there would be mounting evidence for declines in the numbers and diversity of insects, including the pollinators that sustain our food supply.

The second concern, militarism, was very much alive in 1970. At the time, the US was still prosecuting the Vietnam War. There is an essay in *Patient Earth* by Arthur Galston on the use of defoliating herbicides in Vietnam to open up its forests to US bombers,[9] and a primer on radioactivity, addressing both nuclear weapons and nuclear power, which we wrote with Joseph Ginocchio.[10] At the time, avoiding nuclear war was the primary objective among physicists like us who engaged with public affairs. It still ought to be. We had blind spots, of course. We never made the connection between climate refugees and war, nor did we consider oil fields as potential military targets.

The third concern — population — was discussed in practically every environmental textbook in 1970. *Patient Earth* has an appendix on demography (by us), an essay on population by Alice Taylor Day and Lincoln Day,[11] and an essay by Richard Lamm about one of the first state-level initiatives in the US (in Colorado) to loosen the restrictions on abortion.[12] Today, "environment" has distanced itself from "population" in most discourse. Yet, the global population has doubled in the past fifty years and is still climbing, greatly complicating many environmental problems and their solutions. An inexcusable number of women and men still have unwanted children because they have no access to contraception and are unable to exercise freedom over their own reproduction. If *Earth 2020* had included an essay surveying critical population issues over the past fifty years, it would probably have noted that *Patient Earth*, and almost everything written about population in the 1970s, underestimated the demographic transition that would unfold over the subsequent half-century. Today, populations are falling in some countries, and a critical question with environmental significance is whether a similar downward trend will emerge worldwide. If that happens, the global population will decrease, and our species will have an easier time accommodating to this small, shared planet.

In 2020, these three previous concerns have been replaced by two new ones: planetary-scale thinking and environmental justice. We emphasized the first in *Patient Earth*, but to the second we were deaf.

Although *Patient Earth* deliberately focused on US issues in its case studies, again and again it zoomed outward to treat the planet as a whole. We presented the Earth as a single system that could be overwhelmed by human activity in ways that resemble anthropogenic impacts on lakes and airsheds. We taught the reader to perform calculations relevant to global warming, and observed that 'it is ominous that our capacity to change our planet has outrun our understanding of what is happening'.[18] We couldn't have anticipated an ozone hole driven by chlorofluorocarbons (CFCs), but we could have come close; the effect of supersonic airplane emissions on stratospheric ozone was already a live issue.

We did not deal with ocean acidification adequately. We described how the oceans had taken up a portion of anthropogenic CO_2 up to 1970, and commented, briefly, on the increasing acidity of surface ocean waters. We explained chemical buffering, and how increasing the ocean's acidity reduces its capacity to take up more CO_2. But we utterly failed to point out that an increase in acidity was a threat to the ecological integrity of the oceans. We didn't ignore warnings about ocean acidification because there were none then, but we also didn't listen to our own words and pursue their consequences.

The essay about resource scarcity by Charlotte Alber Price — on helium conservation programs — adopted an entirely US perspective.[14] We wrote nothing about world hunger, or ice, or sea level or the world's forests and fisheries — all treated in *Earth 2020*, which is globally-focused throughout. Both books are silent on the overuse of antibiotics, and uncontrollable epidemics — topics that must also be brought into the discussion.

Much of the planetary thinking in *Patient Earth* is at the societal level. Herman Daly, at our invitation, contributed an essay that was the first publication of his path-breaking ideas about 'the Equilibrium Society,' where material flows through an economy reach a plateau.[15] Such zero-growth arguments remain unfashionable (and incomprehensible to economists) today, in about the same way as they did fifty years ago. That essay was complemented by a contribution from Richard Falk on the need to strengthen the international institutions managing the global environmental commons,[16] an argument that is at least as relevant now as then.

Patient Earth did not have a single essay on environmental justice, and, fifty years later, neither does *Earth 2020*. Yet, morally and politically, both within and between countries, inequality and equity are dominant issues. Living and working in New Haven, Connecticut, in 1970, we were surrounded by the symptoms of injustice. Poverty was acute in the city, with the worst local air quality and the major disruptive traffic arteries in the poor neighborhoods. Racial environmental injustice accompanied income-based environmental injustice. The closest we came to addressing this issue in *Patient Earth* was in an essay by H. Lyle Stotts, an emergency room doctor in Bridgeport, Connecticut, who, single-handedly and without community support, was bandaging urban sores.[17] We included the essay to provide an example of what the individual, working alone, can accomplish, but failed to draw a wider circle to include the systemic issue of environmental injustice.

Our light treatment of the intersection between poverty and environment was a consequence of our focus on the environmental problems generated by high consumption. The dominant perspective in *Patient Earth* is that the rich are overconsuming, and the dominant objective from the environmental perspective is to 'decouple' (a word introduced around that time) growth in well-being from growth in material flows. Overconsumption was then, and still is, a dissonant idea.

Both *Patient Earth* and *Earth 2020* emphasize pollution. In 1970, people described the two components of environmentalism as the green and the brown. The green is the protection of unspoiled areas; the brown is the repair of spoiled areas. *Patient Earth* includes not only the already cited essays on herbicides and radiation, but also Alfred Eipper's essay on the overheating of a lake by a nuclear power plant,[18] another by Austin Heller and Edward Ferrand on sulfur dioxide emissions from burning coal,[19] and a third by Orie Loucks on the effort to ban dichlorodiphenyltrichloroethane (DDT) in the US.[20] *Earth 2020* discusses plastics, space junk and contaminants in general. We are glad that plastics have an essay, and that it includes micro-plastics. We could not have anticipated the damage to wildlife caused by these fine plastic particles, a huge problem already today and growing ever larger; photographs of the plastics in the gut contents of wild animals are becoming hard to ignore. There is even credible evidence that these plastic particles move from our food and drinking water to our brains.

Patient Earth did not anticipate endocrine disruptors. Yet, the subsequent brilliant work of Theo Colborn and others on hormone-imitating synthetic chemicals in the environment uncovered a major threat to the health of humanity. One could say that we anticipated this issue, because Galston's essay on herbicide use in Vietnam includes discussion of its teratogenic effects, while Loucks's essay on DDT explains how DDT-induced enzymes produce estrogen breakdown.

*P*atient Earth* was driven by a three-component model of social change: science-policy-activism. Underlying *Earth 2020*, we infer, is the same model, but it is not prominent. The three components work together, not sequentially. The science is well-enough understood to enable the problem and its potential solutions to be identified. The activists use the science to scope the problem, to reduce surprise, and to critique solutions. The solutions require innovations in policy that activists formulate and governments enact. Indeed, the years immediately after 1970 featured a burst of innovative legislation in the US and elsewhere addressing air and water pollution, toxic chemicals and endangered species. Also at that time came legal requirement to evaluate environmental impacts.

The concept of 'well-enough understood science' is a loaded one. Scientists will always want more information, and there are numerous puzzles in any field of science to keep its practitioners busy. But when is the science sufficient for taking action? We have looked back at the progress on the various issues raised in *Patient Earth*, from climate change to biodiversity, from toxics to reproductive freedom, and from warfare to economic sustainability. In each case, we asked whether there was sufficient science in 1970 to know whether action in the form of public policy was needed. We concluded that, yes, the science was generally sufficient to impel such action. Also, the activists' level of awareness was generally high. But the conceptualization of, and commitment to, effective policy was woefully lacking. The imbalance is about the same today.

How much the impacts have grown in fifty years! And the tasks have become more challenging too, despite more relevant science and technology, more policy savvy and more social engagement. Two thirds of the entire increase in atmospheric CO_2 concentrations since 'pre-industrial times' has happened since 1970.[21] What will the next fifty years bring? Leaving aside changes in power politics (*Patient Earth* did not anticipate the rise of China or the fall of the Soviet Union), what about our understanding of the natural environment? Many of the authors of the essays in *Earth 2020* end on an optimistic note. We did, as well. Will people be optimistic fifty years from now about the fifty years after that?

The science today is sufficient to justify activism and policy on many problems, but that is not a reason to slow the scientific quest. As we write, perhaps somebody working at a laboratory bench, or sampling soil in a warming tundra bog, or collecting demographic data for an agency, has a new insight. Maybe we will learn to think more about the deteriorating acuity of our senses resulting from our growing addiction to electronic media. Perhaps we will learn that essential microbes in our guts are being poisoned by the pesticides in our diet, or that our immune systems are being compromised by living in overly sterile homes, or that intense heat waves are harming our brains, or that overfishing is affecting the capacity of the oceans to function as a carbon sink.

In our future, we will have new capabilities to modify organisms, thanks to CRISPR and other tools of the biomedical revolution. We will probably be wrestling with an electricity system largely dominated by energy that is not at our beck and call, because of night and clouds and doldrums. We may be dealing again with nuclear power. And we are likely to be sorting out geoengineering — the deliberate modification of the planet for 'human betterment.' Both 'human' and 'betterment' will be vexing issues: not only which countries get to define 'betterment' (not every country wants less warming), but which trade-offs need to be taken into account so as not to debilitate the non-human while attending to the human. Clearly, the broad enterprise of science must continue, as must the active public engagement of concerned scientist-citizens, such as those we featured in the *Patient Earth* case studies.

The title of our book invoked the twin meanings of 'Patient'. We are in a caring relationship to Earth, as a doctor is to a patient. And in 1970, Earth was willing to wait patiently, as we worked through a diagnosis and searched for appropriate treatment. Half a century on, in 2020, Earth is still our patient, but it has become impatient. The two of us, today, hope, but are by no means certain, that there is yet more time. We are not willing to assert Game Over. At every future moment, there will be better and worse choices, and it will matter which are chosen.

Endnotes

1. We are both authors and editors of *Patient Earth*, New York: Holt, Rinehart and Winston, 1971.

2. Paul Sears, 'An empire of dust,' in *Patient Earth*, 2–15, at 2.

3. Jeremy A. Sabloff, 'The collapse of the classic Maya civilization,' in *Patient Earth*, 16–27.

4. John Harte and Robert H. Socolow, 'Lessons from the past,' in *Patient Earth*, 1.

5. Albert Hill and Michael McCloskey, 'Mineral King: Wilderness versus mass recreation in the Sierra,' in *Patient Earth*, 165–80.

6. Kent Shifferd, 'The fight against Project Sanguine,' in *Patient Earth*, 151–63.

7. John Harte and Robert H. Socolow, 'The Everglades: Wilderness versus rampant land development in South Florida,' in *Patient Earth*, 181–202.

8. Ibid., 182.

9. Arthur W. Galston, 'Warfare with herbicides in Vietnam,' in *Patient Earth*, 136–50.

10. John Harte, Robert H. Socolow and Joseph N. Ginocchio, 'Radiation, in *Patient Earth*, 295–320.

11. Alice Taylor Day and Lincoln H. Day, 'Toward an equilibrium population,' in *Patient Earth*, 206–25.

12. Richard D. Lamm, 'Abortion: A case study in legislative reform,' in *Patient Earth*, 58–69.

13. John Harte and Robert H. Socolow, 'Energy,' in *Patient Earth*, 294.

14. Charlotte Alber Price, 'The Helium Conservation Program of the Department of the Interior,' in *Patient Earth*, 70–86.

15. Herman E. Daly, 'Toward a stationary-state economy,' in *Patient Earth*, 226–44.

16. Richard A. Falk, 'Adapting world order to the global ecosystem,' in *Patient Earth*, 245–57.

17. H. Lyle Stotts, 'Window to the city: The emergency room,' in *Patient Earth*, 31–40.

18. Alfred W. Eipper, 'Nuclear power on Cayuga Lake,' in *Patient Earth*, 112–34.

19. Austin Heller and Edward Ferrand, 'Low-sulfur fuels for New York City,' in *Patient Earth*, 42–57.

20. Orie L. Loucks, 'The trial of DDT in Wisconsin,' in *Patient Earth*, 88–111.

21. See also 'Carbon' by David Archer in this volume.

Climate 1970–2020

—

Tapio Schneider

I grew up in Germany in the 1970s and 80s, where I became a competitive cross-country skier in my teenage years. Back then, the sport was popular in the Harz Mountains near my home, and we could count on 120 days per year with snow on the ground. Today, four decades later, skiing in the Harz Mountains has lost its wide appeal. Winters now average just 65 days per year with snow cover, tendency falling.

Meanwhile, in my current home in Los Angeles, the average number of days with temperatures exceeding 32°C has increased from 53 in 1970 to 67 today. This is two extra weeks' worth of very hot days that desiccate California's landscape, priming it for ferocious wildfires, and days that put vulnerable populations at risk — days when children cannot play sports or have school recess outside, when heat-related emergency room visits by outdoor workers soar, and when deaths among the elderly spike because they are susceptible to heat stroke and heat stress-induced heart attacks.

From 1970 to now, global warming has gone from an abstract threat discussed by scientists to a fact that cannot be ignored. It is here. We feel it. We see it.

The global warming we experience now was predicted long ago. In an 1896 paper that marked the birth of modern climate science, Swedish chemist and Nobel Laureate Svante Arrhenius connected rising and falling CO_2 levels to global warming and cooling in an attempt to explain the waxing and waning of ice ages over Earth's history.[1] From earlier

https://doi.org/10.11647/OBP.0193.04

measurements by others, such as the American astronomer Samuel Langley, Arrhenius knew that CO_2 and water vapor are what we now call greenhouse gases: gases that selectively absorb the infrared radiation emitted by heated bodies (the radiation that warms your hand next to a stove or radiator). Arrhenius demonstrated how rising CO_2 levels would lead to warming by trapping heat near Earth's surface. He also recognized that water vapor exerts an important amplifying feedback, since a warmer atmosphere holds more water vapor, which itself is a greenhouse gas that traps heat.

Arrhenius' model was simple, and the measurements he used were inaccurate. Fortuitously, errors from the simplification and in the measurements largely canceled each other, and he was able to get what is now considered not far from the correct result. Arrhenius predicted that doubling atmospheric CO_2 levels would raise Earth's temperature by 5–6°C. But more important than the precise degree of warming Arrhenius predicted was the fundamental physical insight he delivered: there is a close link between greenhouse gas concentrations and global temperatures. In later work, he observed that burning coal could lead to a significant rise in atmospheric CO_2 levels and appreciable global warming within a few centuries to millennia, a prospect entirely desirable from his Nordic vantage point: 'We would then have some right to indulge in the pleasant belief that our descendants, albeit after many generations, might live under a milder sky and in less barren natural surroundings than is our lot at present'.[2]

Arrhenius' insights proved prescient about what the future would hold, though he and generations of scientists after him severely underestimated the rate at which CO_2 would accumulate in the atmosphere and change the climate.

We now know from historic air preserved in bubbles in the ice sheets of Antarctica and Greenland that atmospheric CO_2 levels hovered around 270 ppm for 10,000 years, following the end of the last ice age. By the late 1800s, however, industrial activities began to increase atmospheric CO_2 levels, which reached 295 ppm by the turn of the twentieth century. Modern measurements of atmospheric CO_2 levels were started in the late 1950s by Charles David Keeling from the Scripps Institution of Oceanography and brought an almost immediate surprise: concentrations were rising more rapidly than anticipated, implying that the oceans were taking up less of the CO_2 emitted by human activities than scientists had previously believed.

By the first Earth Day in 1970, CO_2 levels had reached 320 ppm, 20% above pre-industrial levels. The current value, half a century later, is around 415 ppm, more than 50% more than pre-industrial levels.[3] These values imply that we have added about twice as much carbon dioxide to the atmosphere since 1970 as in all of previous human history before. Worldwide emissions of carbon dioxide from all human sources, including fossil fuels and deforestation, have steadily climbed from 20 billion metric tons per year in 1970 to 42 billion tons now, with no peak in sight. Today, the average North American loads the atmosphere every year with an amount of carbon dioxide weighing about the same as ten midsize-passenger cars. We are releasing CO_2 into the atmosphere far more rapidly than Arrhenius could have possibly imagined.

Along with a growing global network of CO_2 measurements, we have also amassed a large instrumental record of temperature measurements from the nineteenth into the twentieth centuries. In the late 1930s, English engineer Guy Callendar first demonstrated a global warming trend, which he linked to the 10% rise in CO_2 levels that had already occurred by that date. Modern temperature data compiled from all over the world have demonstrated that the average land temperature has increased by 1.4°C since 1900.[4] The vast majority of this increase (1.2°C) has happened since 1970, with a rate of increase in the Arctic (2°C since 1970) that is almost twice the global average. These seemingly small temperature increases hide large changes, leading, in the case of the Arctic, to thawing permafrost and the collapse of structures built on formerly frozen ground.

In response to this warming, the Arctic's summer sea ice cover has plummeted 40% and is approaching its demise.[5] Arctic summers without sea ice will soon be a reality, with enormous implications for human livelihoods and regional ecology.[6] Across the globe, increasing temperatures are associated with a wide range of climate concerns, including stronger rain storms, prolonged droughts and sea level rise.[7]

Even worse, we have yet to see the full extent of the warming to which we have already committed our planet. At least some of the warming associated with increased greenhouse gas levels is masked by air pollution. Over much of the middle to late twentieth century, smog blanketed industrialized areas such as London, Los Angeles and Central and Eastern Europe.[8] Smog consists of tiny aerosol particles, which reflect sunlight back to space, shading and cooling Earth. The added aerosol particles can also increase the number of droplets and ice crystals in clouds, which increases their reflectivity and adds to the cooling effect of air pollution.

Although air quality in the west has improved over the past fifty years (thanks to amendments to the Clean Air Act in the US in 1970, and similar legislation in other western countries that followed), air pollution has worsened in much of the rapidly industrializing world, especially in China and India. The persistence of smog in Earth's atmosphere has thus masked some of the warming that rising greenhouse gas levels otherwise would have caused. As countries improve their air quality, the cooling effects of smog will be reduced, leading to more warming.

Today, we know there's more to climate change and the ways it affects humans than how greenhouse gases regulate the transfer of radiation through the atmosphere. Other processes are also important, including changes in cloud cover, effects of air pollution on clouds, uptake of heat by turbulent ocean circulations and uptake of CO_2 by the ocean and land biosphere. Understanding this complex web of interlinked processes requires more than the calculations Arrhenius performed by hand — it requires computer models.

The first computer-based global climate models were developed in the 1960s and 1970s by pioneers Joseph Smagorinsky and Syukuro Manabe at the US Government's Geophysical Fluid Dynamics Laboratory, Akio Arakawa and Yale Mintz at the University of California, Los Angeles, and Warren Washington and Akira Kasahara at the National Center for Atmospheric Research in Boulder. From these early beginnings more than half a century ago, climate models have steadily become more complex, tracking the exponential increase in computer performance since then.

Current climate models follow the path of solar radiation through the atmosphere to the surface, accounting for what is reflected back out into space and what is absorbed by Earth's atmosphere and surface. They calculate how the heated atmosphere and surface emit thermal infrared radiation, how the radiative heating and cooling drive the motion of the atmosphere and ocean and how air and water transport energy from low to high latitudes, cooling the tropics, warming the poles and enabling life as we know it. Capturing the full complexity of the atmosphere alone is a daunting task, even without including the oceans, biosphere and frozen cryosphere. It is a task far beyond the capabilities of the largest supercomputers today or those of the foreseeable future. Describing just the turbulent motions of the atmosphere requires around 10^{22} numbers characterizing temperature, velocity and humidity at different locations — about the number of molecules in a computer chip, and far beyond what a computer can hold in memory.

To approach the monumental challenge of simulating a coupled Earth system, climate models break down the complexity of the system into coarser chunks. This is achieved by dividing the globe into a grid and then performing computations separately for each box of the grid. The size of the grid's boxes — the resolution at which the model can view Earth — controls the accuracy of its calculations. Early climate models in the 1970s had a grid size of about a thousand kilometers, meaning that a slice across the Atlantic Ocean might span just four or five boxes. Current models with much smaller grid sizes can resolve processes down to scales of tens of kilometers. The most sophisticated models today capture radiative processes and larger-scale turbulence in the atmosphere and oceans, and they include models of the land and ocean biosphere. They have allowed us to explore complex processes, such as the link between global warming and intensification of rainstorms.

But despite significant advances in climate models since the 1970s, some critical processes remain difficult to resolve. The small-scale turbulence that sustains clouds, and processes occurring on tiny scales, such as the microphysical processes shaping droplets and ice crystals in clouds, cannot be accurately represented in current models. Yet even these small-scale processes matter for climate. A cloudy night is warmer than a clear night because clouds are good absorbers of Earth's emitted infrared radiation. Clouds can also make for a cool day at the beach because they reflect sunlight back to space, shading Earth. These small-scale processes affect the trajectory of longer-term climate change, and therein lies the rub — without resolving these processes in climate models, it is difficult to predict precisely how much more warming, extreme storms and sea level rise we should plan for, even if we know how much greenhouse gases will be emitted.

Despite the uncertainty of climate predictions, some things are clear. If greenhouse gas emissions were immediately cut to zero, the level of these gases in the atmosphere would stabilize, before starting a slow decline to a new baseline level over centuries to millennia. But the air would also be cleared of the polluting and cooling aerosols produced by fossil-fuel burning. The result would be more warming in the short term, despite stabilization of greenhouse gas levels. The climate effects of air pollution have not been precisely quantified, but current models suggest that we would see an additional global average warming of 0.4–1.7°C within years of eliminating all greenhouse gas emissions.[9]

We cannot stop CO_2 emissions suddenly; our energy economy has the agility of an oil tanker. Over the past fifty years, growth in global energy demand has outpaced growth in energy production from renewables. Greenhouse gas emissions are growing with no peak in sight, much less a reduction to zero. There is virtually no chance that we can avoid the 1.5°C global-average warming above pre-industrial temperatures aimed for by the Paris Agreement in 2015 (signed in 2016).[10] If we consider the 1.1°C global-average warming that has already occurred since the nineteenth century, and the time-delays in our energy economy and in the climate system, the inescapable conclusion is that we are on track to exceed 1.5°C and perhaps even 2°C global-average warming above pre-industrial temperatures.

While not physically impossible, limiting global warming to 1.5°C requires an implausibly short-term turnaround of greenhouse gas emissions, and staying within a 2°C warming target requires an economic restructuring at a pace not previously seen in history. Just to have a fighting chance of avoiding more than 2°C warming, we would have to drop greenhouse gas emissions down to zero within about 30–40 years — the lifetime of today's fossil-fuel power plants. Even achieving zero emissions in that timeframe would give us only a two-thirds chance of limiting global warming to 2°C above pre-industrial levels, according to the generation of climate models that came out in the early 2010s.[11]

Worse still, many of the most recent climate models are running hotter, indicating a higher sensitivity of the climate system to greenhouse gases than previously considered likely. This result stems in part from recent findings that the cooling effect of polluting aerosols may be stronger than previously thought. But if cooling by air pollution in the past was stronger than previously estimated, the sensitivity of the climate system to increases in greenhouse gases must be larger than previously estimated, or else we would not be able to account for the twentieth-century temperature rise. If the new models are more accurate than the previous generation — which is unclear at present — we may have underestimated the warming response to greenhouse gases. In that case, limiting global warming to 2°C above pre-industrial levels will be extremely challenging, if not impossible.

From the first Earth Day in 1970 to today, global warming has moved from an abstract scientific prediction to a reality we must contend with. At the same time, the discussion of global warming has moved from an exclusive focus on mitigation to the deepening realization that adaption is also critical. Mitigation was the focus of the 1992 United Nations Framework Convention on Climate Change, in which countries around the world committed to 'stabilize greenhouse gas concentrations in the atmosphere at a level that would prevent dangerous anthropogenic interference with the climate system'.[12] Follow-on treaties such as the Kyoto Protocol (1997) and the Paris Agreement were attempts to make this specific and enforceable. What 'dangerous anthropogenic interference with the climate system' means remains unclear. Nonetheless, we do have an idea of where we are headed.

The last time carbon dioxide levels were sustained at today's levels (around 415 ppm) was three million years ago, during the mid-Pliocene. At that time, Earth's global mean temperature was 2–3°C warmer than today, and sea level was about 17 m higher. The Greenland ice sheet was ephemeral and the Antarctic ice sheets were smaller; the water locked up in them now was part of the oceans. Mammalian life on Earth was thriving, but Homo sapiens did not yet exist, and neither did currently low-lying cities such as Alexandria, Amsterdam, Cape Town, Guangzhou, London, Miami, Mumbai, New York, Osaka, Rio de Janeiro or Shanghai. Even today's greenhouse gas levels, if sustained for centuries, must be considered dangerous for human civilizations that are adapted to the relatively stable climate and coast lines that existed for the 10,000 years before the industrial revolution.

Mitigating global warming to the greatest extent possible remains essential to prevent the cataclysms that await when current greenhouse gas levels are sustained for centuries, or increase even further. After decades of failures, efforts to stem rising tides and temperatures are much more urgent now than in 1970 or 1992, when snow and skiing were still common in the Harz Mountains of my childhood. But mitigation alone no longer suffices. Climate change will leave no one untouched. We have no choice but to adapt.

Endnotes

1. For a reprint of Arrhenius' paper and a discussion of its context and reception, see D. Archer and R. Pierrehumbert (eds.), *The Warming Papers: The Scientific Foundation for the Climate Change Forecast*, Chichester, UK: Wiley-Blackwell, 2011, https://doi.org/10.1002/met.1289

2. Quoted in H. Rodhe, R. Charlson and E. Crawford, 'Svante Arrhenius and the greenhouse effect', *Ambio*, 1997, 26, 2–5, at 4.

3. For historical and current concentrations of CO_2 in the atmosphere, see https://www.esrl.noaa.gov/gmd/ccgg/trends/

4. For a modern compilation of Earth temperature data on which these figures are based, see http://berkeleyearth.org

5. For sea ice data, see https://nsidc.org/cryosphere/sotc/sea_ice.html

6. See also 'Ice' by Julian Dowdeswell in this volume.

7. See also 'Weather' by Neville Nicholls and 'Sea Level Rise, 1970–2070: A View from the Future' by Robert E. Kopp in this volume.

8. See also 'Air' by Jon Abbatt in this volume.

9. The short-term committed warming estimates here indicate the warming that results from eliminating aerosols and short-lived greenhouse gases associated with fossil-fuel burning. The estimates are based on the simple model put forward by T. Mauritsen and R. Pincus ('Committed warming inferred from observation,' *Nature Climate Change*, 2017, 7, 652–55, https://doi.org/10.1038/nclimate3357), updated using recent modeling results. For more discussion of the challenges posed by committed warming, see V. Ramanathan and Y. Feng ('On avoiding dangerous anthropogenic interference with the climate system: Formidable challenges ahead', *Proceedings of the National Academy of Sciences of the United States of America*, 2008, 105, 14245–50, https://doi.org/10.1073/pnas.0803838105).

10. Available at https://unfccc.int/resource/docs/2015/cop21/eng/l09r01.pdf

11. These estimates are based on the carbon budgets in Table 2.2 of R. K. Pachauri and L. A. Meyer, eds., *Climate Change 2014: Synthesis Report. Contribution of Working Groups I, II and III to the Fifth Assessment Report of the Intergovernmental Panel on Climate Change*, Geneva: IPCC, 2014, https://www.ipcc.ch/report/ar5/syr/). They include cumulative carbon dioxide emissions since the reference year in the report and assume an immediately starting and approximately linear ramp down of emissions to zero.

12. Available at https://unfccc.int/resource/docs/convkp/conveng.pdf

Politics and Law

—

Elizabeth May

I was in tenth grade on 22 April, 1970. The announcement of a full day of actions and teach-ins reached me — if memory serves — some months earlier. United States Senator Gaylord Nelson launched the call, mirroring the grassroots tactic of the campus teach-ins against the Vietnam War. In that time before the internet, I must have heard about Earth Day through my membership in Friends of the Earth and the Sierra Club. I had time to plan a door-to-door canvas of the community, recruiting other high school friends and obtaining permission to miss school (unlike the Swedish activist Greta Thunberg, who initiated a global school strike movement for climate action). I had time to write up an information sheet about the environmental threats of the day — over-fertilization of water ways from phosphates in detergents (eutrophication), nuclear bomb tests, air pollution and water contamination. No such thing as printing documents at home in those days. I typed up the facts and asked my father, who had an office and a secretary, if he could get 500 photocopies printed for me. I had to pay for them from my allowance.

Some things about that first Earth Day have been lost to the mists of time. The essence was counter-culture, yet the establishment and corporate world readily embraced the event in an attempt to greenwash themselves. US Republican President Richard Nixon issued a proclamation endorsing the day's event, planting a commemorative tree on the White

    https://doi.org/10.11647/OBP.0193.05

House South Lawn. Coca-Cola and the chemical manufacturing giant DuPont also signed up to celebrate the Earth. Over the past half-century, Earth Day has become a ritualized, 'safe' event, layered with hypocrisy and opportunism. Yet, this event still persists in raising awareness and action, and its first celebration, in 1970, was a landmark in many ways.

For one thing, Earth Day 1970 launched the United Nations into planning the first international conference on the threats to our environment — the 1972 Stockholm Conference on the Human Environment. It also propelled many governments to create — for the first time — Departments of Environment. The US Environmental Protection Agency (EPA) was set up within months of Earth Day on July 9, 1970, and similar government agencies were soon established in industrial countries around the world, charged with creating and enforcing new environmental legislation. Sweeping new amendments to the US Clean Air Act and Clean Water Act came in effect in 1970, and 1972. At the same time, the world's first environmentally-focused political party, the Popular Movement for the Environment (PME), was founded in 1972 in the Swiss canton of Neuchâtel. Less than a decade later, the PME's Daniel Brélaz, a mathematician by training, became a member of the Swiss national parliament. These were heady days for the burgeoning environmental movement.

Looking back over the five decades of environmental law and policy since Earth Day 1970, what stands out? There are two threads to follow in the emerging challenge of environmental governance: the global North–South divide (1970–1990) and the emergence of global corporate rule (1990 onwards). The latter is particularly important, as it threatens to undo hard-won progress in preserving Earth's natural systems.

For the first two decades after Earth Day, all industrialized countries started down the path of controlling pollution with a focus on science-based regulations and policy. Issues confronted in this period were largely solved at local and regional scales: eutrophication, acid rain, local air quality, visible water pollution from factories and sewers, and so on. These problems were both created and solved within the national context of wealthy industrialized countries. There was no need for diplomacy or multi-lateral negotiation.[1] Nor was there an apparent need to recognize the uneven burdens and responsibilities of global environmental degradation.

The vast majority of developing countries were absent from the 1972 UN Stockholm Conference. They collectively and deliberately boycotted what was decried as the wealthy countries' agenda. The only prominent developing country leader to attend was Indian Prime Minister Indira Gandhi. Virtually all of the other developing country governments slammed the conference. Pollution was not seen as a priority for nations unable to feed their people.

A major global effort to bridge this apparent North–South divide took the form of the World Commission on Environment and Development. The commission was chaired by the sitting Prime Minister of Norway, Gro Harlem Brundtland, and its membership included retired heads of government and leading figures from both the industrialized and developing world. Its 1987 report, *Our Common Future* (also known as the Brundtland Report),[2] attempted to overcome the pollution versus poverty argument by embracing a new concept of sustainable development. This idea was grounded in a fundamental principal of equity, both in the present and for future generations. The report's authors called for human society to take actions in three key areas — environment, development and militarism. They also called for a major UN Summit to take place on the twentieth anniversary of the UN Stockholm Conference. By 1989, the UN General Assembly voted to hold a major Summit in June 1992, where Environment and Development would be the focus. Population was set aside for the next major gathering for Women's Issues in Beijing, and militarism was dropped altogether.

Around the time of the Brundtland Report, the old North–South divide surfaced in the negotiations to protect the ozone layer. At that time, in the late 1980s, science was increasingly demonstrating the massive danger from chlorofluorocarbons (CFCs), a group of 'miracle compounds' used in a variety of household products from aerosol cans to refrigerators. Initially believed to be benign and indestructible, it became clear that CFCs were breaking down in the stratosphere, releasing reactive chlorine atoms that gobbled up molecules of ozone.[3] Stratospheric ozone plays a critical role in screening out the sun's most harmful ultraviolet rays, and the loss of this protective layer was a cause for significant concern. The massive ozone hole that developed over Antarctica became emblematic of this threat, and helped spur the world to action.

As industrialized countries mobilized to develop a protocol leading to an effective ban on ozone-depleting substances, the developing world expressed concern that, once again, this was an issue for the rich countries of the world. In the Global South, rotting food was a bigger issue than thinning ozone, and developing countries wanted to expand their use of CFCs, particularly a class of these compounds known as freons for refrigeration. Even more uncomfortable was the reality that the skin cancer threat was highest for the pale Caucasian inhabitants of industrialized countries in North America and Europe.

The CFC negotiations were very difficult and protracted. The solution eventually emerged in September, 1987, during negotiations in Montreal. The key breakthrough was the acceptance that industrialized and developing countries had to be treated differently, under a novel principle of 'common but differentiated responsibilities' (CBDR). This new principle allowed all developing and industrialized countries to sign on to legally binding requirements. The industrialized countries committed to immediately begin reductions of ozone-depleting substances to 50%, while developing countries could increase their use by 15%. All parties to the convention agreed to base their actions on science and modify their targets as the science required. Critically, they also agreed to enforce the targets through trade sanctions. Within a few years, both industrialized and developing countries were on board for the total elimination of ozone depleting substances. To this day, the Montreal Protocol remains one of the greatest success stories of the environmental movement.[4]

Five years later, in 1992, the United Nations convened the Rio Earth Summit, with the goal of burying the North–South divide. Along with most developing nations, Brazil had boycotted the 1972 UN Stockholm Conference. But now, Rio hosted what was to that date the world's largest global summit. Every government on Earth attended — with Fidel Castro and George H. W. Bush Sr. sharing the same stage. The so-called 'Rio Bargain' led to the creation of major treaties aimed at protecting the environment, alongside targets to eliminate poverty and increase well-being in the Global South.

The Rio Earth Summit was viewed as a huge success. The commitment to transfer wealth and technology from North to South was bundled in non-binding commitments under the so-called Agenda 21.[5] In contrast, the commitments to preserve biodiversity and avert the climate crisis were subject to binding treaties (the Biodiversity Convention,[6] and the UN Framework Convention on Climate Change, UNFCCC).[7] On the surface, these binding agreements represented milestones in international environmental cooperation. But they lacked enforcement mechanisms, and this would prove to be a fatal flaw.

The 1990s dawned one year after the Berlin Wall crumbled in 1989. Soon after, the USSR disintegrated and the Cold War ended. All our hopes were directed toward the long-awaited peace dividend. In most industrialized countries, the largest part of government budgets had been squandered on one of the world's largest polluters — the military. Activists had pressed for years for military budgets to shrink and to redirect finances to the elimination of poverty. Now was the chance. Canadian scientist and environmental champion Dr. David Suzuki dubbed the 1990s the 'turnaround decade'. And why not? A major UN Summit had delivered plans to focus on environmental protection and poverty reduction, while at the same time, the Cold War and its massive waste of resources on the military was over.

Unfortunately, the heady optimism of the Rio Earth Summit was blunted almost immediately. In July, 1992, the most powerful industrialized nations convened at the G-7 Summit in Munich. The discussion of this group virtually ignored the Rio Earth Summit and its commitments to the developing world. The Munich G-7 communiqué did reference the work of the Rio Earth Summit, but the thrust of the commitment shifted to multi-lateral trade negotiations and economic growth.[8]

Perhaps more than any other development, it was the creation of the World Trade Organization (WTO) that ushered in a new world economic order and turned the 1990s from the 'turnaround decade' to the 'could-have-been decade'. The WTO was built on the post-war framework of the General Agreement on Tariffs and Trade (GATT), greatly expanding the scope of international trade and shifting the framework to increased corporate rights. While the GATT had focused on reducing barriers to trade of goods, the web of agreements in the Uruguay Round of WTO negotiations dealt with far more than goods. It set out global rules for intellectual property, services and a shift to corporate rights in trade agreements.

The North American Free Trade Agreement (NAFTA) also came into force in this period, further entrenching corporate rights in trade negotiations. For example, Chapter 11 of NAFTA created — for the first time — the right for foreign corporations to seek damages when government action (including environmental regulation) reduced their expectation of profits. Compensation could, for instance, be based on government actions to restrict the production and storage of toxic chemicals near drinking water. In the language of NAFTA, such government efforts were 'tantamount to expropriation'.[9]

As the relative power of transnational corporations increased under the protection of NAFTA, the rights of the nation state for environmental protection and other social benefits declined. Indeed, the WTO committee on Trade and the Environment was tasked not with examining potentially harmful environmental effects of trade, but with identifying environmental treaties that posed barriers to international trade. This approach set the stage for the emasculation of environmental treaties as global corporate rule spread through an expanding web of investor-state agreements.

What is astonishing is the degree to which the ascendency of the WTO was accepted without question. The GATT, upon which the WTO was built, had never set out such sweeping powers for corporate profit rights to trump governmental jurisdiction. Indeed, Article XX of the GATT[10] created provisions to support government policy measures deemed necessary to protect human, animal or plant life, or the conservation of finite natural resources. These provisions still exist, but they have been all but ignored by the WTO.

In 1997, shortly after the WTO was established, the Kyoto Protocol was negotiated in Japan to protect climate stability.[11] Ten years earlier, the successful Montreal Protocol had used trade sanctions as an enforcement mechanism. In Kyoto, however, the trade ministers of industrialized countries instructed the environment ministers that trade sanctions were not an option. Even Canada, which had led the way in the fight against CFCs, significantly changed its stance. The Canadian Environment Minister, Christine Stewart, went to Kyoto with a clear message that her country would not sign the agreement if trade sanctions were included. This sentiment was echoed by many other wealthy nations, with the result that the Kyoto Protocol was left with no effective enforcement mechanism. Like the 1992 UN Biodiversity Convention, and the UNFCCC, Kyoto had no teeth.

Far too often, journalists and observers examining the failure of humanity to respond to the climate crisis overlook the rise of corporate power as a dominant factor. Other excuses — scientific uncertainty, high costs or uncertain benefits — are given weight. But what really happened was clear to those of us who were eye witnesses. As noted by the late Jim MacNeill, chief author of the Brundtland Report, the 1992 Earth Summit in Rio marked the beginning of the 'carbon club'. Large transnational fossil fuel giants saw the threat. They knew it was existential. The threat they registered was not the fact that their product threatened human life on earth; it was a threat to their industry and its profits.

As Naomi Klein wrote in her 2014 book *This Changes Everything*, the climate crisis has experienced the problem of 'bad timing'.[12] Just as the world began to see compelling evidence of human impacts on the climate system, the rise of the WTO and the growing international reach of corporate power acted to limit meaningful action to address climate change. Some of the major fossil fuel giants began spending hundreds of millions of dollars to mislead the public. Their strategy — taken from the playbook of the big tobacco lobby — was (and is) to create doubt about science.[13] And they have been successful. As a result, we failed to take appropriate action when we had the chance to avoid the climate emergency we now experience. But we still have the chance to avert the worst. The worst is nearly unthinkable — so we push it to the back of our minds. The worst is crossing a point of no return, where human-caused greenhouse gas emissions trigger unstoppable, self-accelerating global warming.

My own transition in the last part of this chronology has been from an activist in non-partisan civil society to an actor on the stage of partisan politics. While all those working to preserve a healthy biosphere — activists, academics, industry leaders or elected politicians — play an important role, my own path has led me to seek desperately needed change through involvement in Green Party politics.

From their early humble beginnings in the 1970s, Green Parties now exist in nearly ninety countries around the world. These parties have exerted significant political influence as members of coalition governments, most notably in Germany, where they have been instrumental in the phase out of nuclear power. Green Parties come in many flavors, but they all adhere to fundamental values, including environmentalism, social justice and non-violence. Greens recognize the need to simultaneously address the two large trends underpinning the deterioration of Earth's natural systems — the North–South divide and global corporate rule. We believe that achieving the seventeen UN Sustainable Development Goals will ensure both climate action and social justice.[14]

As we strive for a truly sustainable future, we must deconstruct global corporate rule and put the global survival ahead of corporate greed. It is time to put large fossil fuel companies on notice. Governments around the world must be prepared to revoke the corporate charter of any company that threatens the integrity of a habitable biosphere. We can and should create a World Trade and Climate Organization to ensure both prosperity and survival.

We have had fifty years of experience in environmental law and policy. And from that we know that we are failing our own children and the myriad of other species with which we co-inhabit this planet. Yet solutions are available. We already have the tools we need to confront the climate emergency. The very same tools that work for liberalization of trade and the protection of intellectual property can work to deliver global climate action, ensure sustainable economic development and eliminate poverty.

Why not?

Endnotes

1. See also 'Climate Negotiation' by Rosemary Lyster in this volume.

2. World Commission on Environment and Development, *Our Common Future*, Oxford: Oxford University Press, 1987, http://www.un-documents.net/our-common-future.pdf

3. See also 'Air' by Jon Abbatt in this volume.

4. Available at https://ozone.unep.org/treaties/montreal-protocol-substances-deplete-ozone-layer/text

5. Available at http://sustainabledevelopment.un.org/content/documents/Agenda21.pdf

6. Available at https://www.cbd.int/convention/text/

7. Available at https://unfccc.int/process-and-meetings/the-convention/what-is-the-united-nations-framework-convention-on-climate-change

8. Available at http://www.g8.utoronto.ca/summit/1992munich/communique/index.html

9. Available at https://www.international.gc.ca/trade-agreements-accords-commerciaux/topics-domaines/disp-diff/nafta.aspx?lang=eng

10. Available at https://www.wto.org/english/tratop_e/envir_e/envt_rules_exceptions_e.htm

11. Available at https://unfccc.int/resource/docs/convkp/kpeng.pdf

12. N. Klein, *This Changes Everything: Capitalism vs the Climate*, New York: Simon & Schuster, 2014. See especially 'Part one: Bad timing', 26–164

13. See also 'Media' by Candis Callison in this volume.

14. Available at https://sustainabledevelopment.un.org/?menu=1300

Carbon

—

David Archer

Carbon is the backbone of all life on Earth. This element is able to accommodate up to four molecular bonds, giving it great chemical versatility and the ability to assemble into a wide variety of molecules, from sugars, fats and proteins (the building blocks of life), to the complex hydrocarbons that fueled the industrial revolution. By weight, carbon makes up only about 0.03% of our planet, yet this element exerts a profound influence on virtually every aspect of the Earth System. And perhaps more than any other element, carbon has been the subject of intensive debate over humanity's impact on the global environment.

On Earth, carbon is partitioned among a number of different reservoirs, including the crust and mantle (99.95%), dissolved and particulate forms of inorganic carbon (0.049%), living organic material in the terrestrial and marine biospheres (0.00064%), and atmospheric trace gases (another 0.00064%), including carbon dioxide (CO_2) and methane (CH_4). These 'greenhouse' gases absorb out-going infra-red radiation from Earth's surface, trapping heat within the planetary system like a thermal blanket. Of all the forms of carbon, most recent attention has been focused on CO_2, whose atmospheric concentrations have been significantly altered by human activities, with profound impacts on Earth's climate. Understanding how the concentration of atmospheric CO_2 is controlled requires an

https://doi.org/10.11647/OBP.0193.06

appreciation of complex processes that act to regulate the distribution of carbon among its different global reservoirs.

On long-term geologic time scales, carbon flowing into and out of the solid Earth (primarily continental and oceanic crust) acts to stabilize climate, in a negative feedback loop known as the CO_2 weathering thermostat.[1] When dissolved in water (including rain), carbon dioxide acts as an acid, reacting with continental igneous rocks to release minerals that are eventually transported to the oceans, where they regulate seawater chemistry and pH. The calcium that is released from this 'weathering' process reacts with dissolved carbon in seawater to produce the mineral calcium carbonate ($CaCO_3$), which sinks through the water column and is buried in ocean sediments for hundreds of thousands of years or longer. This burial of carbonate minerals is the main pathway for pulling carbon out of the atmosphere and storing it in long-term geological reservoirs. The rate of this carbon storage process depends on the climate (temperature and especially rainfall), which itself depends on the concentration of CO_2 in the air. As atmospheric CO_2 concentrations increase, so too does the rate of CO_2 removal through weathering reactions. Hence the thermostat.

The co-evolution of Earth's climate and biosphere has not always gone entirely smoothly.[2] The weathering CO_2 thermostat sometimes lurches toward warmer or colder set points for a period, challenging the biosphere to adapt. Take, for example, those episodes in Earth history when huge floods of volcanic lava released world-changing amounts of CO_2 into the atmosphere. Volcanic gases become greatly enriched in CO_2 when hot magma rapidly heats sedimentary rocks, causing them to explode with CO_2, methane and other gases. In the present day, a large fraction of Earth's volcanic CO_2 emissions comes from just a few volcanoes, which are mostly located in the tropics and associated with sedimentary calcium carbonate deposits. But the CO_2 emissions from these modern volcanos are tiny compared to the massive volcanic sources of the distant geological past. The largest of the mass extinctions, at the end of the Permian period 250 million years ago, was driven by one of the largest volcanic floods in Earth history, in present-day Siberia.[3] This huge CO_2 release to the atmosphere overwhelmed the capacity of the CO_2 thermostat to adjust, leading to a large spike in global temperatures that radically changed environmental conditions on Earth, resulting in wide-spread species extinction.

Sometimes, the biosphere itself lurches suddenly in a new direction, impacting the carbon cycle and global climate. For example, at the end of the Devonian period, about 350 million years ago, plants began to colonize Earth's land surface, with the evolution of roots, leaves and seeds that allowed them to extract water from the ground and disperse their offspring. These early terrestrial plants enhanced the weathering reaction on land by chemically attacking the rocks and forming soils as a by-product. The faster rates of weathering also removed a huge amount of CO_2 from the atmosphere, sending the planet into an ice age. Moreover, the colonization of land by plants led to a massive release of phosphorus into the ocean, which fertilized marine algae, resulting in oxygen depletion and extinction in the deep ocean.[4]

Fast forward several hundred million years or so, and CO_2 still seems to be the mother of all environmental problems.[5] Humans first exerted a significant impact on the atmospheric CO_2 concentration by clearing land for agriculture or game management. A century of deforestation in North America and Europe, from 1800–1900, caused the land surface to become a source of carbon to the atmosphere. Around the same time, humanity's first substantial use of fossil carbon arose with the invention of the steam engine by James Watt, giving our species the means to generate mechanical power on a large scale. Invention and ingenuity took hold quickly, pushing forward an unprecedented revolution of technology that transformed life on Earth over a mere two centuries. In a geological heartbeat, humanity consumed vast amounts of organic carbon deposits that had formed over hundreds of millions of years.

Since 1750, humans have released about 330 billion metric tons of carbon. Approximately half of all these emissions have occurred over the last half-century, since the first Earth Day in 1970. Additional human impacts on the carbon cycle have come from continued land use changes and cement production at massive scales (cement fabrication can be considered as a 'reverse weathering' process that liberates CO_2). While the land surface of Europe and North America may now, fortuitously, be re-absorbing CO_2 through the regrowth of trees, deforestation in other regions continues to provide a source of CO_2 to the atmosphere.[6] The future of the land carbon pool depends significantly on human land use practices, but also on the stability of huge deposits of frozen organic carbon in northern permafrost soils. Warming tundra and Arctic soils are accelerating the melting of these frozen deposits, which will likely release more CO_2 than any other part of the land surface could match.[7] Taken together, these human perturbations of the global carbon cycle are analogous to the volcanic CO_2 releases in the 'greenhouse extinctions' of the geological past. The total quantities of CO_2 liberated naturally by volcanos were probably larger than humans could muster by burning fossil fuels, but the rate of our CO_2 emissions are likely unprecedented in Earth history.

What happens to all of the CO_2 released by human activities? About half of it is still in the atmosphere, with the concentration rising from around 320 parts per million (ppm) in 1970 to around 415 ppm today (a roughly 30% increase). The rest of the anthropogenic carbon has mostly been absorbed into a giant oceanic pool, which has helped to stabilize both the atmospheric CO_2 concentration and the temperature of Earth's surface (and thus global climate). Over the past fifty years alone, the oceans have absorbed about 150 billion metric tons of CO_2 from the atmosphere, while also absorbing significant amounts of heat. In the short-term, oceanic uptake of CO_2 and heat are mitigating the greenhouse effect. Over the longer-term, however, CO_2 and heat pollution stored in the ocean will eventually be re-released to the atmosphere, slowing down any future recovery. In addition, CO_2 uptake by the oceans has a significant effect on seawater chemistry, resulting in increasing acidity (decreased pH) as hydrated CO_2 becomes carbonic acid. The global-scale response of the ocean carbon cycle to a shift toward greater acidity is difficult to predict, but we do know that ocean acidification was a prominent feature of previous mass extinction events on Earth.

The time it takes for ocean pH to recover from an abrupt increase in CO_2 concentrations is on the order of thousands of years — long by human standards, but short geologically. And herein lies a critical distinction between human-derived fossil fuel carbon and natural volcanic CO_2 sources. Whereas volcanic CO_2 was released into the atmosphere over millions of years, fossil-fuel carbon has been released over the last couple of centuries at a rate that overwhelms the capacity of natural chemical buffering processes. Our rapid CO_2 emissions will thus lead to larger spike in ocean acidity than any previous disturbance of the carbon cycle.[8]

Political negotiation on the climate issue has focused on trying to limit peak global temperature rise to 1.5°C.[9] This much warming would make the planet warmer than it has been in millions of years, since long before the development of humans as a civilized species. Warming of 2°C or more would almost certainly be worse, but the choice of 1.5°C itself is somewhat arbitrary. A true 'safety' boundary could be defined in terms of the energy balance of the planet. Today, due to the rising CO_2 concentration, the amount of solar energy delivered to Earth from sunlight, exceeds the energy lost from the planet. This energy imbalance is causing the planet to warm, with most of the excess heat going into the ocean. The concentration of CO_2 in the air that would balance Earth's energy budget, and thereby stop the buildup of this heat pollution, is about 350 ppm. This threshold was crossed about thirty years ago. The current atmospheric CO_2 concentration, 410 ppm, is rising by a few ppm per year.

Even if human CO_2 emission stopped today, cold turkey, the CO_2 concentration in the air would remain above 350 ppm for thousands of years; essentially forever, from our perspective. Engineering a return to a stable, optimal climate state may thus require actively removing CO_2 from the atmosphere. There are multiple possible strategies for doing this, including stimulating the growth of plants (we would have to bury the resulting carbon), using chemical scrubbers (as is done on submarines and spaceships), and artificially increasing weathering rates (by grinding up certain kinds of rocks that react with CO_2).[10] But whatever approach we take, limiting atmospheric CO_2 levels will be extremely difficult and costly. Getting back to 350 ppm within a few decades would require removing about 440 billion metric tons of carbon from the atmosphere. Optimistically, if it costs $360 to remove one metric ton of carbon from the atmosphere,[11] the total bill would be $160 trillion, about 1.6 year's worth of global world economic activity.

In Earth history, innovations such as the development of mining — whether by dirt-eating worms, rock-cracking roots, or fracking oil drillers — are able to upend Earth's chemical metabolism and alter its climate. Today, humanity is gorging on the energy of fossil fuels, eating the fat of the land, like a giant mold thriving on an old crust of bread. In a world of biological opportunism and growth, the conclusion would be foregone: exponential growth of the consumer population followed by collapse when the nourishment is gone. However, of all of the climate episodes and extinctions in the history of the carbon cycle, this is the first in which the agent of the event is at least beginning to *understand* the consequences of its actions.

Our perturbation of the carbon cycle is primarily an energy problem, so fundamental to our lives that it is challenging to imagine changing it quickly enough. But there is plenty of energy all around us, from the sun, and in the wind. If we were simply running out of fossil fuel now, would our civilization really collapse? Much of the human activity on planet Earth is driven and guided by our financial system; when there is immediate money to be made, we are extremely clever and adaptable.

Fossil CO_2 can be seen as a waste-management problem, like that of Shel Silverstein's Sarah Cynthia Sylvia Stout, who would not take the garbage out.[12] The poem describes how Sarah's house filled up with all manner of solid waste. If Sarah's consumption habits were typical of a North American child, it might take a few years for her house to fill completely with garbage. By comparison, the mass of invisible waste CO_2 from her fossil energy use would be about thirty times that of the visible solid waste. If her CO_2 waste also remained in the house, it would flush out all of the air within about a month, killing Sarah like a gawping fish.

Before the Great Stink in 1858, the sewers of London emptied directly into the Thames River. Massive overhauls of the rudimentary sewer system must have been controversial at the time, but business as usual was no longer an option. Neither is it a viable option now, as we come to understand that our waste CO_2 is not so different from the chamber pots of Victorian Londoners. The challenge lies in making the decision. The global scope of CO_2 emissions means that everyone has to cooperate in the eventual solution, even if the benefits of cleaning up are far less immediate to individuals. It is a question of ethics versus finance, analogous to the institution of slavery, which has been largely eliminated multiple times in human history. In many ways, things are going in the right direction, with costs of carbon-free energy becoming competitive with existing coal power, for example. At present, however, our progress — driven by our money-oriented decision-making system — is too slow.

Endnotes

1. J. C. G. Walker, P. B. Hays and J. F. Kasting, 'A negative feedback mechanism for the long-term stabilization of Earth's surface temperature', *Journal of Geophysical Research*, 1981, 86, 9776–82, https://doi.org/10.1029/JC086iC10p09776

2. P. Brannen, *Ends of the World: Volcanic Apocalypses, Lethal Ocean and Our Quest to Understand Earth's Past Mass Extinctions*, New York: Harper Collins Publishers, 2017.

3. S. Z. Shen et al., 'Calibrating the end-Permian mass extinction', *Science*, 2011, 334, 1367–72, https://doi.org/10.1126/science.1213454

4. T. J. Algeo and S. E. Scheckler, 'Terrestrial-marine teleconnections in the Devonian: links between the evolution of land plants, weathering processes, and marine anoxic events', *Philosophical Transactions of the Royal Society B-Biological Sciences*, 1998, 353, 113–28, https://doi.org/10.1098/rstb.1998.0195

5. P. P. Tans, 'An accounting of the observed increase in oceanic and atmospheric CO2 and an outlook for the future', *Oceanography*, 2009, 22, 26–36, https://doi.org/10.5670/oceanog.2009.94

6. On deforestation, reforestation and afforestation, see 'Forests' by Sally N. Aitken in this volume.

7. S. M. Natali et al., 'Large loss of CO^2 in winter observed across the northern permafrost region', *Nature Climate Change*, 2019, 9, 852–57, https://doi.org/10.1038/s41558-019-0592-8

8. K. Caldeira and M. E. Wickett, 'Anthropogenic carbon and ocean pH.', *Nature*, 2003, 425, 365, https://doi.org/10.1038/425365a. On ocean acidification and warming, see also 'Oceans 2020' by David M. Karl in this volume.

9. H. J. Schellnhuber, S. Rahmstorf and R. Winkelmann, 'Commentary: Why the right climate target was agreed in Paris', *Nature Climate Change*, 2016, 6, 649–53, https://doi.org/10.1038/nclimate3013

10. On carbon dioxide removal (CDR) and negative emissions technologies, see 'Geoengineering' by D. G. MacMartin and K. L. Ricke in this volume.

11. D. W. Keith, G. Holmes, D. St. Angelo and K. Heidel, 'A process for capturing CO2 from the atmosphere', *Joule*, 2018, 2, 1573–94, https://doi.org/10.1016/j.joule.2018.05.006

12. S. Silverstein, 'Sarah Cynthia Sylvia Stout Would Not Take the Garbage Out', in *Where the Sidewalk Ends*, New York: Harper Collins Publishers, 2014, 70–71.

Everyday Biodiversity

—

Jeffrey R. Smith and Gretchen C. Daily

For many living in the urban century, waking up to a raucous dawn chorus of birds is a near-unimaginable possibility. The shift from birdsong to the alarm clock (and now the smart phone) emblemizes the dramatic transformation of human experience of nature throughout our daily rituals. It underscores how accustomed we've become to the synthetic world we've created, and our growing alienation from the declining biodiversity around us — with wild bird populations in the United States and Canada having dropped by nearly 30% since the 1970s.[1] This is a case of 'shifting baseline syndrome', where we acclimatize to a new 'normal', failing to recognize the ongoing ecological tragedy that is unfolding around us. We go about our daily routines without thinking about the multitude of ways in which biodiversity enriches our lives, what its continued loss implies for our future well-being and how we can intervene to slow, and hopefully reverse, the dramatic global declines of nature in its variety and abundance.

If you're a coffee drinker, your very first sip in the morning connects you to an incredibly complex web of interactions between plants, animals, fungi and the biophysical systems that support them. Coffee, like tea and other domesticated crops, was once an unremarkable plant fighting for survival among thousands of other species. In its native range, it had to compete with other plants for water, sunlight and nutrients, while avoiding being eaten by insects, browsing mammals, fruit-loving birds and the like. It was this constant struggle against potential herbivores that started an evolutionary arms race

https://doi.org/10.11647/OBP.0193.07

that led to the development of caffeine (and many other culinary delights) as a defensive compound. This evolutionary arms race drove genetic changes that created the plants we know today as *Cafe arabica* and *Cafe robusta*. Through selective breeding, we have further altered coffee biodiversity at its most basic genetic level to improve yield and quality, creating cultivars that are drought and disease resistant, and more desirable in a host of other ways. This process highlights a fundamental attribute of biodiversity; it encapsulates all levels of biological organization from the genes that make up individual species to the ecosystems that support them.

Let's turn our attention back to breakfast, considering the bowl of fresh fruit, jam spread across toast, or orange juice you might have alongside your morning coffee. Almost certainly, these fruits will have relied on pollination carried out by a bee, moth, fly, beetle, hummingbird, bat, or some other living thing. In fact, over 75% of the vitamins and nutrients we consume come from crops with animal pollinators, and our most valuable and nutritious — and most delicious — crops are, by and large, dependent on these creatures.[2] But many of these pollinators are in trouble. There has been a rapid increase in morality in managed honeybee hives, accompanied by widespread reductions in native pollinator abundance and massive declines in insect abundance generally.[3] The reasons for these declines are complex and not fully known, but likely include land-use and climate changes, pesticide use and other forms of pollution.

The global decline of pollinators is symptomatic of a much larger global trend. Since the dawn of the industrial era, species extinction rates have accelerated dramatically. Today, we are losing an estimated 1,000 to 10,000 times more species per year than would be natural under pre-human conditions.[4] And the surviving species are dwindling rapidly, with about 60% of wild vertebrate populations — amphibians, reptiles, mammals, fish and birds — shown to be in decline.[5] Today, in 2020, the total weight ('biomass') of humans and livestock is estimated to be twenty-five times larger than that of all remaining wild mammals.[6]

The trends of declining biodiversity are troubling but not mysterious. We understand the root causes. Earth's wild plants, animals and other life forms are in decline because of overhunting and overharvesting, converting habitats into ever-expanding agricultural land and cities, and by a litany of indirect impacts, including spreading invasive species, pollution and now, increasingly, climate change. These impacts are not new. Human activity appears to have driven species extinctions for nearly 10,000 years, with much of the megafauna of the North American Continent disappearing concurrently with the arrival of the first human beings. The intensity of these pressures has only increased as our population, per-capita consumption, and technological prowess have grown. These human impacts on biodiversity became particularly notable after the Industrial Revolution when, for example, demand for feathers for the millinery trade drove the passenger pigeon to extinction. Once the United States' most ubiquitous bird, it occurred in flocks of billions that famously could take days to fly over a town.

We have reckoned with dozens of high-profile species at risk of extinction. The thin line between species survival and extinction is perhaps no more evident than in the story of the bald eagle, the national symbol of the United States. Driven to perilously low levels in the 1970s by overuse of the insecticide dichlorodiphenyltrichloroethane (DDT) and other toxic chemicals, this majestic bird presented a true crisis for American conservation, and American society more broadly. Thanks to the Endangered Species Act and other crucial legislation prompted by the first Earth Day in 1970, we have managed to recover bald eagle populations to healthy levels. In fact, the species was removed from the Endangered Species List in 2007, as its populations had sufficiently recovered to viable levels.

While the rescue of the bald eagle is only one success story, it illuminates a pathway for saving the hundreds of other species currently under federal protection. But developing sound management practices becomes more challenging, though not impossible, as we increase the scale of the factors driving species loss, the scope of the species considered, or the complexity of stakeholder relationships. These are the challenges we must face in dealing with climate change impacts on Arctic sea ice and polar bear populations, universal declines in North American grassland birds, or the impacts of wolf reintroduction into the intermountain Western United States. The successes of species-centric conservation from 1970 to today will surely guide the next half-century of conservation efforts to stave off a looming mass extinction. But one thing is certain; the sooner we act to prevent species declines, the more successful we will be. Moreover, swift action now is likely to save us an immense amount of resources, financial and otherwise, that we would need to invest down the road to achieve the same results.

The loss of biodiversity is not just a matter of disappearing species, but also of radical landscape transformation. There is perhaps no starker example of such transformation than the rise of urban areas around the globe. While accounting for less than 5% of Earth's area, cities now house almost 60% of the human population.[7] Even prior to the rise of 'mega-cities', many of the earliest conservation movements around the world were based on the separation of human-dominated systems and wilderness areas. Beginning in the late nineteenth century with the creation of Yellowstone National Park, protected areas have played a central role in conservation. The basic idea of this approach is to establish areas for nature to thrive beyond human pressures. Research has shown that, while no place is untouched by the hands of humanity, such wilderness preserved are, indeed, essential for reducing the extinction risk of species. The designation of protected areas has only intensified in the wake of the first Earth Day, and we see campaigns for the augmentation of protected areas, with calls for 30% protection by the year 2030.

Yet protected areas will never be enough — increasingly, they are islands, too small, too few, and too remote to support the biodiversity upon which human society depends. This was perhaps one of the most visionary turns of the Environmental Movement of the 1970s. No longer was US conservation focused only upon protected areas; rather, the passage of the Endangered Species Act, the Clean Water Act and the US Clean Air Act underscored the need for biodiversity to be protected in the sea of humanity. This was done not only for the inherent value of biodiversity, but the realization that our own species depends on functioning ecosystems to provide vital life-support services. In cities, for example, green spaces and street trees reduce temperatures in urban heat-islands, purify urban air and attenuate city noise. Moreover, daily exposure to such natural elements has been shown to have manifold benefits to mood, attention span, and memory retention over standard urban or suburban landscapes.[8] And across sweeping landscapes and seascapes, ecosystems produce important goods (such as timber and seafood), essential life-support processes (such as natural pollination and water purification), life-fulfilling conditions (such as beauty, serenity and inspiration) and preservation of potential future benefits (for example, the conservation of genetic diversity for future use in agriculture or medicine).

Although government action plays a major role in biodiversity conservation, we must also employ other tools going forward. Increasingly, this means engaging with the economic system to create more ecologically-sustainable goods. Consider the refreshing beer you might drink at the end of your long workday. The global beer industry has been long dominated by a few key players. However, with the recent insurgence of small craft microbreweries, the rules of the game are changing quickly. Many consumers are now willing to pay a premium for beer that boasts both a greater flavor profile and a greater corporate sustainability ethos. This sustainability is achieved through a variety of approaches, including the use of spent hops as agricultural feed (rather than sending them to landfill), and partnerships with local conservation groups to secure forests situated upstream of key water supplies. While it is true that some of these actions are being taken to improve corporate 'green' image of the company ('greenwashing'),[9] sustainable business practices are now not only possible, but increasingly profitable. At the same time, we are seeing increasing public scrutiny and boycotts of companies and industries that refuse to incorporate the value of biodiversity into their decision-making. One of the most prominent examples is the refusal of many consumers to buy products containing palm oil — a crop whose rapid proliferation is endangering tropical rain forests around the globe.

Governments at various levels are also increasingly taking the economic value of nature into account. New York City became a posterchild for this movement in 1997 when it opted to secure its drinking water quality by investing in natural capital rather than building a physical treatment plant. The decision was based on economic analysis, showing a capital cost of $6–8 billion for building a water treatment plant, plus annual operating expenses of $300 million, as compared to an estimated $1–1.5 billion, in perpetuity, for habitat protection in the source watersheds about 100 miles north of the city.[10] Twenty years of experience show that the natural capital investment is working, yielding a triple win — safe water for the ten million people living in New York, compensation for a public service long supplied by farmers and foresters upstream, and protection of many other benefits under the umbrella of safe drinking water. Over the past two decades, this case inspired adoption of similar projects by over fifty major cities in Latin America, a rapidly growing number in Asia, and some now in Africa. Globally, an estimated 25% of major cities stand to benefit from this approach.[11]

Also in 1997, Costa Rica adopted national economic incentives for biodiversity conversation, pioneering a payments for services (PES) scheme that incentivized local farmers to conserve or restore their forests in recognition of the economic returns from increased eco-tourism, carbon storage and water purification (for hydropower — a major export — as well as for irrigation and drinking).[12] This proved to be the beginning of a global trend, with many countries soon establishing similar programs. For example, China launched their own PES program in 1999, enrolling 120 million households in restoring steeply sloping lands for flood protection and water purification.[13] Today, there are over 550 such programs around the globe, with total annual payments of nearly $40 billion.[14] We may thus be witnessing the beginning of a new paradigm, where global economic systems have begun to account for natural capital in order to make wise and sustainable decisions.

As your day finishes, you may find yourself sitting in your back yard or strolling through a local park, enjoying a small vestige of our natural world. It is easy to despair the global decline in nature over the past half-century, yet we can still draw inspiration from the beauty of biodiversity, beyond all of the benefits it provides us in tangible and quantifiable ways. Take a look around as dusk turns to night and you might be fortunate to see some fireflies or lightning bugs. Birds and butterflies have shown us color arrangements that we could only hope to see in the paintings of great masters, while vistas such as the Grand Canyon or the Swiss Alps remind us of the enormity of the world, and ourselves as mere actors in an unfolding play. Planet Earth's species, habitats, ecosystems and landscapes are fundamental to who we are as human beings. We've evolved among them, and have come to appreciate their nuance and beauty in a way that is irreplaceable with the constructs of human hands. If we want to return to a world where waking up to birds singing is the norm rather than a Hollywood fantasy, and where the next generation has a chance of enjoying similar levels of security and well-being that we experience, we must take bold action, and we must do so quickly.

Endnotes

1. K. V. Rosenberg et al., 'Decline of the North American avifauna', *Science*, 2019, 366, 120–24, https://science.sciencemag.org/content/366/6461/120, https://doi.org/10.1126/science.aaw1313

2. A.-M. Klein, B. E. Vaissière, J. E. Cane, I. Steffan-Dewenter, S. A. Cunningham, C. Kremen and T. Tscharntke, 'Importance of pollinators in changing landscapes for world crops', *Proceedings of the Royal Society B: Biological Sciences*, 2007, 274, 303–13, https://doi.org/10.1098/rspb.2006.3721

3. D. Goulson, 'The insect apocalypse, and why it matters', *Current Biology*, 2019, 29, R967–R971, https://www.cell.com/current-biology/pdf/S0960-9822(19)30796-1.pdf, https://doi.org/10.1016/j.cub.2019.06.069

4. J. M. D. Vos, L. N. Joppa, J. L. Gittleman, P. R. Stephens and S. L. Pimm, 'Estimating the normal background rate of species extinction', *Conservation Biology*, 2015, 29, 452–62, https://doi.org/10.1111/cobi.12380

5. Word Wildlife Fund & Zoological Society of London, *Living Planet Index*, 2018, http://www.livingplanetindex.org/projects?main_page_project=LivingPlanetReport&home_flag=1

6. Y. M. Bar-On, R. Phillips and R. Milo, 'The biomass distribution on Earth', *PNAS*, 2018, 115, 6506–11, https://doi.org/10.1073/pnas.1711842115

7. United Nations Department of Economic and Social Affairs, *Revision of World Urbanization Prospects*, 2018, https://www.un.org/development/desa/publications/2018-revision-of-world-urbanization-prospects.html

8. G. N. Bratman et al., 'Nature and mental health: An ecosystem service perspective', *Science Advances*, 2019, 5, eaax0903, https://doi.org/10.1126/sciadv.aax0903

9. See also 'Politics and Law' by Elizabeth May in this volume.

10. G. C. Daily and K. Ellison, *The New Economy of Nature: The Quest to Make Conservation Profitable*, Washington, DC: Island Press, 2002.

11. J. Salzman, G. Bennett, N. Carroll, A. Goldstein and M. Jenkins, 'The global status and trends of Payments for Ecosystem Services', *Nature Sustainability*, 2018, 1, 136–44, https://www.nature.com/articles/s41893-018-0033-0.pdf, https://doi.org/10.1038/s41893-018-0033-0

12. FONAFIFO and MINAET, *Costa Rica. Tropical Forests: A Motor for Green Growth*, San José: FONAFIFO and MINAET, 2012, https://www.fonafifo.go.cr/media/1554/cr-tropical-forests.pdf

13. Z. Ouyang et al., 'Improvements in ecosystem services from investments in natural capital', *Science*, 2016, 352, 1455–59, https://science.sciencemag.org/content/352/6292/1455, https://doi.org/10.1126/science.aaf2295

14. Salzman et al., 'The global status and trends', 2018, 136.

Energy

—

Elizabeth J. Wilson and Elias Grove Nielsen

Homer's epic poem from the eighth century BC recounts the legend of Odysseus' return to Ithaca after the Trojan War. What should have been a journey of a few weeks across the Aegean Sea, became, in Homer's tale, a ten-year ordeal plagued by natural hazards, monsters and divine malfeasance. Imagine the different outcome if Odysseus had owned a diesel outboard motor, a GPS to plot his route through the islands, an emergency radio to track storms and uncooperative winds, and an echo sounder to avoid submerged rocks (and mermaids). And with a cell phone, he could have called his wife Penelope to let her know he'd been delayed, keeping her lurking suitors at bay. Today, Google Maps charts Odysseus' trip from modern day Turkey to Greece as taking less than twenty-four hours, more like a weekend road trip than an Odyssey.

There can be no doubt that modern energy has transformed how humans move, eat, live and play, while also radically altering our impact on Planet Earth. Over the past 10,000 years, new power-producing technologies have been the foundation of modern societies. In human history, today is the energy anomaly. Supported by energy, more people live longer now than in any other time in the history of our species, with access to vastly improved healthcare, sanitation and seemingly limitless opportunities for travel. Energy has also benefited social mobility; no longer are 70–90% of humans serving as serfs and slaves needed to farm and transport goods; in many countries, traditional 'women's work'

  https://doi.org/10.11647/OBP.0193.08

of cooking, laundry and cleaning has been drastically cut by energy-driven appliances (and with men sharing the work!).

But whenever there is progress, there is also regress and unanticipated consequences. Both coal mining and natural gas and oil production, including refining and combustion, affect land use and pollute water and air. Uranium is mined to produce nuclear-powered electricity, creating long-lived radio-active waste. Hydropower entails the damming of rivers, which results in the flooding of tracts of land and impacts upstream and downstream habitat, as well as water flow and quality. Even renewable energy sources, like wind power turbines or solar photovoltaics, require energy and rare Earth minerals in their construction. This 'embedded energy' can be traced through all consumer goods. Energy production and generation also impact human health; respiratory difficulties from bad air quality affect people living near industrial and energy-producing facilities. The World Health Organization estimates that 4.2 million people annually die prematurely from poor air quality.[1] And we should not forget the social costs of energy extraction, transport and use, which range from civil unrest around the location of energy production facilities and pipelines, to corruption, fraud, human rights abuses and large-scale geopolitical engagements. Energy is the largest business on the planet. With energy comes power, both literally and figuratively.

Sometimes, energy production systems experience catastrophic failures. Take, for example, the nuclear accidents at Three Mile Island, Chernobyl and Fukushima Daiichi, oil spills of the Exxon Valdez and BP Deepwater Horizon; or natural gas pipeline explosions in San Bruno, California or Andover, Massachusetts. Such disasters have enormous environmental impacts, and they can also leave a long-term political legacy, as was the case for the Santa Barbara oil spill. On January 28, 1969, an oil platform blowout in the Santa Barbara Channel released three million gallons of oil into the sea floor, creating a massive oil slick on the surface ocean that lead to the death of thousands of birds, fish and marine mammals. Widespread public outcry following this event (the largest oil spill in US history at the time), ushered in sweeping new environmental legislation, and galvanized a burgeoning political and social movement that culminated with the first Earth Day, just one year later.

Over the past half-century since the Santa Barbara oil spill, our energy dependence has grown significantly. Global energy consumption has increased by roughly 45% per capita since 1970, with an accompanying rise in global atmospheric CO_2 concentrations from around 320 parts per million (ppm) in 1970 to over 410 ppm in 2019 (with the annual mean in 2020 likely coming close to 415 ppm). The trend appears to be continuing unabated; in 2018, the world used 3% more energy than the previous year, with accompanying annual CO_2 emissions increasing at 2% to 37 billion metric tons of carbon dioxide.[2] The fear of a climate tipping point looms large, and energy system decarbonization is now more critical than ever.

The good news is that there has been significant progress on the transition to alternative energy sources. Economic incentives, including tax credits, feed-in-tariffs and other subsidies, have stimulated research, development, deployment and investment, and helped reduce the costs of renewable energy globally. As a result, the addition of low carbon energy production capacity has surpassed expert predictions in scale and speed. Innovation and global investment in renewables like solar and wind power topped $300 billion per year for the fifth year running, and 2018 saw near record numbers of renewables, and lower carbon natural gas dominate new energy installations.

The bad news is that our progress has not been nearly enough. Overall global growth in energy demand is outstripping decarbonization efforts, and fossil fuel consumption continues to increase as more people are using more energy around the planet. Today, roughly 80% of energy used still comes from fossil fuels including coal oil and natural gas. This is down from 94% in 1970, but the absolute increase in global energy demand means greater total emissions.

Researchers often discuss 'energy transitions', examining past societal shifts from wood to fossil fuels; or charting future courses from high carbon fossil fuels to low-carbon futures. While these transitions can be locally transformative, global energy transitions have not been substitutive, but additive. While the EU and North America have transitioned from wood as the primary energy source, global wood consumption remains as high as ever. This underscores the fact that the deployment of new energy technologies remains local, shaped by regional priorities and resulting policies. In other words, energy systems are more than just a collection of coordinated technologies, they enshrine social practices and values. There is no *one* global energy system; energy is not distributed, delivered and used equally around the world.

As a privileged citizen of an industrialized country, my experience of energy is vastly different from most of the world's population. When I wake up, shut off the alarm clock, turn on the light, start the hot shower and the coffee maker, and get cold milk from the refrigerator, energy use is almost invisible. When I flip on the light switch, I expect light. This is the privilege of the energy rich, the roughly 2.2 billion people of the 7.7 billion on the planet today who have the luxury of not having to think about energy. For these people, there is more than enough energy for basic comforts, health, food, transportation and wellbeing. There is enough for them to travel by airplane and car, to use cell phones and have Jacuzzis, extra freezers and nose-hair trimmers. This is not to say that energy use is uniform even within energy rich countries; energy disparities do exist, and some citizens in these countries still experience energy poverty.

Abundant energy has enabled rapid economic growth of industrialized societies, and this development has been responsible for the bulk of historical greenhouse gas emissions. But today, energy demand in many rich countries is flat to declining. In 2018, the twenty-eight European Union countries had flat or negative energy consumption growth due to policies supporting increased system efficiencies, investments in renewables and a mild winter. While reducing greenhouse gas emissions has become a modern rallying cry for the energy rich, others often have different priorities.

At the other end of the spectrum are the energy poor, the 1.1 billion people who live without access to electricity; and an additional 2.5 billion without access to modern cooking fuels. In these societies, lighting is often provided by candles or kerosene lanterns, while wood, dung, or charcoal provide energy for cooking and heating. Cooking over a three-stone fire requires significant amounts of both wood and time. Fire is dangerous, and it also creates smoky particulate matter which causes respiratory and eyesight problems, mostly in women and children who do the bulk of global wood gathering and cooking. For these people, whose meagre energy use has not contributed to global climate change, affordability and access to energy is paramount.

In the energy middle, you find 4.4 billion global citizens with access to modern sources of energy, but with varying degrees of reliability and affordability. Here, flicking on a light switch does not guarantee illumination. Until recently, residents of Katmandu, Nepal, were connected to the electric grid, but struggled to get sufficient power.[3] The city provided schedules of when different neighborhood residents could expect to have electricity, and many residents invested in backup solar and inverter systems. However, recent engineering and political reforms have changed this situation, and now residents have a reliable power supply, and only some industries are without power for four hours a day. Now, small businesses no longer have need for their expensive and polluting diesel generators for backup power, and they have enough electricity to run machinery and expand production. Instead of being invisible, energy access drives and shapes personal activities and economic growth.

Globally, energy demand, and associated greenhouse gas emissions, are growing most rapidly in the energy middle. Since 2000, per capita energy demand in China and India has grown by roughly 250% and 50%, respectively, and these development-driven trends continue. In just one year, 2018, energy demand grew by about 8% in India and 4% in China, as compared to a decrease of 0.6% in the European Union. Increasing energy demands in the developing world come with environmental and health impacts. In big cities like New Delhi or Cairo, air pollution is at record levels and urban residents suffer the consequences.[4]

While local and regional impacts of energy use can be managed if political will, technological acumen and economic investments align, addressing a changing global climate requires coordinated global action. The Paris Agreement, drafted in 2015 and signed in 2016, with nearly 200 signatory countries representing 89% of global emissions, constituted a start at collective action for a collective problem.[5] But so far, only two countries (Morocco and The Gambia) are on track to meet their <1.5°C Paris commitments, highlighting the challenges of transforming and adapting legacy energy systems.[6] The International Energy Agency estimates that current investments in renewable and clean energy resources must increase from $900 billion in 2018 to $2.3 trillion per year to meet the Paris Agreement's aggressive greenhouse gas reduction targets, while providing the energy needs for the planet's projected population of almost 10 billion by 2050.[7] The required transformation goes far beyond building new energy production plants and battery storage capacity; rather we require a systematic change in how we use energy — deep efficiency — and how we shift demand to accommodate significant use of variable renewable resources like wind and solar.

As we build our energy systems to reduce greenhouse gas emissions, we must also adapt them to the realities of climate change. Rising sea levels, intensified storms and stronger hurricanes threaten communities and their energy infrastructures. Current energy systems were not designed to withstand the 200 mph windstorm gusts, massive wildfires, floods, or 20-foot tidal surges. As record strength typhoons, cyclones and hurricanes batter vulnerable landscapes in Asia, Africa and the Americas, the fragility and criticality of energy infrastructures are underscored. When the power goes out, gas pumps and credit cards no longer work, cell phone service goes down and streetlights go dark. At home, food in refrigerators begins to rot, water no longer flows and electric heat and cooling systems stop working, leaving people vulnerable to extreme temperatures and disease outbreaks. Here again, the impacts fall disproportionately on the world's poorer populations. When Hurricane Maria hit Puerto Rico in 2017, over 1,000 people died in the aftermath. After Cyclones Idai and Kenneth hit Mozambique six weeks apart in March and April 2019, the risk of waterborne diseases like cholera was a major part of the emergency response.

Humanity now faces the daunting challenges of creating future energy systems that can both mitigate and adapt to changing climates across countries with different economic realities. Many communities and utilities are already adapting to new climate vulnerabilities by replacing wooden power poles with concrete and installing new meters and switches to allow grid operators to better detect and respond to power outages.[8] For example, more resilient cables and flood-proof equipment, coupled with the relocation of substations out of flood zones can enhance the resilience of core energy systems. The use of advanced technologies, including drones, is now helping to remotely assess and monitor damage from storms and ensure more rapid recovery. Some utilities are building in system redundancy, upgrading distribution networks, and investing in micro-grids to provide energy to critical infrastructure. But planning for novel risks is difficult. New patterns of floods, droughts, fires and other natural disasters heighten societal vulnerabilities and force communities to relocate and fortify energy and other human infrastructures.

Today, people live with dramatically different levels of energy access and use; yet we face a common threat in climate change. The energy rich, the energy poor and those in-between have different needs, risks and responsibilities. Responding to this planetary-scale threat requires simultaneous reduction of greenhouse gas emissions to near-zero, and adaptation of infrastructure to emerging (though uncertain) climate risks. All of this, while still providing energy access to a rapidly growing global population. No pressure.

Our future depends on how we will collectively make and use energy. This will be shaped by how we design, travel and live in our cities, communities and homes. Today, a zero-carbon energy system pushes the limits of technology and faces immense political and economic barriers. With the EU Green Deal goal of a carbon neutral Europe by 2050, and at least €100 billion to support it, this is a critical first step.[9] Whether we like it or not, our energy systems are changing. It remains to be seen how they can be adapted to support our collective futures on Earth.

As we look forward, humility should accompany our energy system transitions. The ancient Greeks believed that hubris led to punishment and suffering. This is the tale of Odysseus, who was punished for his arrogance, and of Prometheus, who stole fire from the Gods as a gift to humanity. Praying to the weather gods will not save us from the next hurricane, fire, or the impacts of a changing climate. We should not lose faith in science or political systems, but we might want an extra set of oars at the ready. They may come in handy when that diesel motor sputters out.

Endnotes

1.   World Health Organization, 'Ambient air pollution: Health impacts', https://www.who.int/airpollution/ambient/health-impacts/en/

2.   M. Muntean, D. Guizzardi, E. Schaaf, M. Crippa, E. Solazzo, J. G. J. Olivier and E. Vignati, *Fossil CO2 emissions of all world countries — 2018 Report*, Luxembourg: Publications Office of the European

Union, 2018, https://ec.europa.eu/jrc/en/publication/fossil-co2-emissions-all-world-countries-2018-report; BP, *Statistical Review of World Energy 2019*, London: BP, 2019, https://www.bp.com/content/dam/bp/business-sites/en/global/corporate/pdfs/energy-economics/statistical-review/bp-stats-review-2019-full-report.pdf

3. B. Sangraula, 'How Nepal got the electricity flowing', *The Christian Science Monitor*, 16 January 2017, https://www.csmonitor.com/World/Asia-South-Central/2017/0116/How-Nepal-got-the-electricity-flowing

4. See also 'Air' by Jon Abbatt in this volume.

5. Available at https://unfccc.int/resource/docs/2015/cop21/eng/l09r01.pdf

6. A. Erickson, 'Few countries are meeting the Paris climate goals. Here are the ones that are', *The Washington Post*, 11 October 2018, https://www.washingtonpost.com/world/2018/10/11/few-countries-are-meeting-paris-climate-goals-here-are-ones-that-are/

7. International Energy Agency, *World Energy Investment Report 2019*, Paris: International Energy Agency, 2019, https://www.iea.org/wei2019/overview/

8. F. Stern, S. Hendel-Blackford, K. Leung, I. T. Rogrigo Leal and D. Vitoff, 'Extreme weather alert: How utilities are adapting to a changing climate', *Utility Dive*, 6 March 2019, https://www.utilitydive.com/news/extreme-weather-alert-how-utilities-are-adapting-to-a-changing-climate/549297/

9. European Commission, 'A European Green Deal: Striving to be the first climate-neutral continent', https://ec.europa.eu/info/strategy/priorities-2019-2024/european-green-deal_en; European Commission, '2050 long-term strategy', https://ec.europa.eu/clima/policies/strategies/2050_en

Forests

—

Sally N. Aitken

In this era of cheap global travel, it is tempting to visit a new place every year to experience diverse landscapes and cultures. But there is also something to be said for observing the same place year after year, slowly and carefully. To understand the dynamic nature of forests, and their response to climate change and other human-induced pressures, it helps to witness the changes firsthand in one place over many years.

That place for me is a small cabin on a remote lake in the Chilcotin region of British Columbia, Canada. Ten hours from Vancouver by car, and two hours from the nearest grocery store, it is set in a landscape of forests and mountains, with just a few hardy souls scratching a living off the land through forestry, ranching, or tourism. Grizzly bears, black bears and moose are common residents, and seasonal visitors include sandhill cranes, pelicans and Arctic terns. Trees in these forests live long, slow lives, growing only a little each summer in preparation for the deep cold of winter.

Since the first Earth Day in 1970, mean annual temperatures in the Chilcotin have risen by about 1.5°C. That may not sound like a lot, but it is more than the global average of 1°C warming over the last fifty years and equal to the average temperature difference between Vancouver, British Columbia and Portland, Oregon, 500 km to the south. One might predict that a little warming would make life easier for trees in such a cold place; instead, the forest is unravelling.

   https://doi.org/10.11647/OBP.0193.09

Freezing winter temperatures historically kept insects such as the mountain pine beetle at bay by periodically killing off a large segment of the population. With milder temperatures and fewer cold spells, beetle populations have exploded over the past twenty years, killing pine trees across eighteen million hectares of forests in British Columbia, an area the size of Washington State. The ravaged Chilcotin forests are now filled with dead, grey trees — some still standing and many covering the forest floor. After the mountain pine beetles came spruce beetles, Douglas-fir beetles and western spruce budworms. Tree defenses were weakened by drought and unable to mount sufficient chemical and physical defenses against attack. The beautiful, white-stemmed aspen are also in decline due to increasing impacts of insects and diseases. To keep a hiking trail clear of fallen trees in these parts is to develop a physical awareness of the extent of tree mortality.

And then came the wildfires. Both 2017 and 2018 were record-shattering years for wildfires in British Columbia, with 2.5 million hectares burned. In early July 2017, my partner and I were forced out of the Chilcotin by a thunderstorm that started over a hundred fires in a single day. The only road out led right through the heart of a wildfire. Although it was still daytime, we drove through dense smoke as dark as night. The only light came from trees alongside us bursting into flames, or stems glowing bright red from bottom to top. This massive fire eventually merged with others to cover 467,000 hectares, the largest ever recorded in the province.

Thankfully, the summer of 2019 was wet and cool in my part of the world. There was fresh snow on the mountains one July morning, perhaps fueling doubts in those who question the reality of climate change. But heat waves in Europe and over much of North America made 2019 the hottest June and July on record globally. People have short memories of weather, and current conditions can distort our perceptions of climate trends. We also have ways of modifying our environments and clothing to suit conditions, insulating us from an external reality. But trees, long-lived and sedentary, must tolerate what comes. Weather events impact tree growth, as recorded in their wood rings over decades, centuries, or even millennia. From these annual growth records, we know that trees can tolerate considerable climate fluctuations. But what happens when those tolerances are exceeded?

Like the trees of the Chilcotin, forests in many parts of the world are suffering due to climate change and other ecological and environmental pressures. Drought-related tree mortality, alone or in combination with insect outbreaks, has been documented on every continent in the past two decades.[1] In California alone, a recent multi-year drought and associated insect outbreaks have caused the death of nearly 150 million trees.[2] While this is just a small percentage of the total number of trees in that state, it is an unsustainable mortality rate for tree species that can live well over a century. In Germany, an estimated one million trees have died in the past two years, generating fears that some parts of the country are becoming unforestable.[3] Drought-related megafires have burned across western North America, Europe and Australia. High northern latitudes, in particular, are experiencing greater warming than elsewhere on Earth. Boreal forests in North America and Siberia are showing those effects. Trees in so-called drunken forests are tilting and tipping as permafrost melts, dying from drought and insects and burning in vast wildfires.

While climate change is impacting the health of the world's forests, humans are also accelerating climate change through deforestation. Since 1970, the total area deforested on Earth has increased by 3 million km², an area comparable to the size of India. Prior to 1900, most deforestation was in temperate regions, but recent clearing has been almost exclusively in the tropics, with just 50% of tropical forests remaining globally,[4] and nearly 20% of Amazonian forests lost over the past half-century. Many tropical forests have been razed to make way for industrial agriculture (e.g., cattle grazing, soy production for animal feed and oil palm plantations). In recent years, there had been hopeful signs that deforestation had slowed. Between 2004 and 2012, for example, Brazil's annual deforestation rate dropped considerably. Sadly, this progress has been reversed as the country's pro-development government turns a blind eye to massive, largely illegal forest clearing and burning. These activities resulted in the Amazonian fires of 2019, when more than 80,000 fires burned across Brazil, capturing global attention (at least for a short while). There is a real fear that further deforestation and burning in the Amazon will alter local climate — increasing temperatures and decreasing rainfall — driving that ecosystem past a tipping point where rainforests will be replaced by arid savannahs. Global attention and pledges of support internationally cannot repair the damage that has already been done.

It is easy to cast blame on lower- and middle-income countries where rapid deforestation is currently underway. But we must remember that inhabitants of the industrialized world live in previously forested cities and towns, and farm previously forested fields. Nearly half of global forests were cleared by humans across history, not just in recent decades. We also consume agricultural products that come from deforested tropical areas, fueling economic drivers of deforestation. And in higher-income countries, some natural forests are still being converted to plantations of non-native species such as eucalypts or pines. Such introduced species significantly alter local ecology, and can bring other risks as well. For example, fire-prone eucalyptus plantations in Portugal exploded in flames in 2017, killing sixty-four people in the deadliest wildfire in that country's history.

The degradation of Earth's forests, and tropical rainforests in particular, presents us with both an ecological and climatic catastrophe. Globally, forests and other terrestrial vegetation absorb approximately two billion metric tons of carbon dioxide annually, and have absorbed just over one quarter of the carbon dioxide released to date by human activities in the Anthropocene.[5] This carbon dioxide is taken up during photosynthesis, and the resulting carbohydrates are used to build leaves, stems, branches and roots (carbon makes up about half of the dry weight of wood). Forest soils also store substantial amounts of carbon in fine roots, organic matter, fungi and microorganisms. Together, forest trees and soils currently store more carbon than all of the readily exploitable oil, gas and coal reserves globally, with about half of this stored carbon in tropical regions.

Harvesting of forests, or burning of trees during forest fires or land clearing, converts them from net sinks to sources of greenhouse gases. If cleared sites are rapidly reforested, the net impact on greenhouse gas emissions is lower than if the land is converted to another use, such as agriculture or urban development. Deforestation is responsible for about 18% of global greenhouse gas emissions — more than the total global emissions from transportation. Forest fires are another significant source of carbon dioxide to the atmosphere. In 2017 and 2018, for example, British Columbia wildfires emitted three times as much carbon dioxide as the annual burning of all fossil fuels in that province. After burned and insect-infested trees die, their decay converts them into a further source of atmospheric carbon dioxide, creating a positive feedback cycle. Fires and insect outbreaks have turned British Columbia's vast forests from a net carbon sink to a net carbon source, contributing to rather than mitigating greenhouse gas emissions.

The world's forests are also critical reservoirs of biodiversity, containing 80% of all terrestrial species globally and about three quarters of birds, with most of these found in the tropics. In North America, for example, the total number of birds has dropped by 29% since 1970, a reduction of about three billion individuals.[6] Climate change is only one of the human factors destroying the library of life; others include overharvesting,[7] pollution,[8] and loss of habitat. A recent UN report concluded that up to one million species are at risk of extinction.[9] To slow or avert this biodiversity crisis, forests must be restored or maintained to provide habitat for the many species they house.

So, what is the solution? Perhaps we can simply plant more trees? Indeed, a large global effort to replenish previously existing forests (reforestation) or create new ones (afforestation) has been proposed as the most realistic and cost-effective climate change strategy. Trees are certainly less expensive and easier to scale up than other greenhouse gas reduction technologies. A recent analysis using satellite imagery concluded that there is room globally for an additional 0.9 billion hectares of continuous forest, representing about 500 billion trees.[10] At first glance, this appears to be a win-win solution, restoring native forests, generating forest-based goods and services for local communities, enhancing greenhouse gas sinks and creating habitat for biodiversity. But this view from space misses many details that will determine the feasibility of this solution on the ground.[11]

Afforesting grasslands won't actually increase carbon sequestration, as we now know that grassland soils contain as much carbon as forests, and they regenerate soil carbon faster than forests. We must also consider the individuals and communities that will benefit or be harmed as a result of reforestation. Per-capita greenhouse gas emissions are highest in high-income countries, while deforestation is greatest in those with lower incomes. Some people living in poverty who are already being disproportionately impacted by climate change may also suffer losses in livelihoods from widespread reforestation through loss of grazing or agricultural land. If local communities are not involved in designing and benefiting from reforestation, tree planting programs are destined to fail.

In some places, adapting landscapes to climate change may require planting fewer, not more, trees. While planting more trees per hectare might result in more carbon sequestration, forest managers are shifting to lower-density forests in drought-prone areas to provide trees with sufficient water and mitigate wildfire risks. Some forests will likely shift to grassland ecosystems with further warming. We need to better understand where and when these ecological shifts will occur, and how they will impact carbon storage.

In places where tree planting offers more benefits than risks, we need to consider climate change when selecting trees to plant. Changes in climate over the lifetime of a typical tree will impact forest health and reduce carbon sequestration. Populations of trees of a given species vary genetically, depending on the climates they have evolved in. Scots pine from Finland and Spain, for example, differ considerably in their growth timing, cold hardiness and drought tolerance. Tree species and populations used today for reforestation will need to be carefully chosen to increase the likelihood that they will be healthy and productive throughout their lifetimes as climates change. We are dealing with a moving climate target, and without a crystal ball to pick the best trees for the uncertain climate 50 or 100 years from now, we should hedge our bets by planting a variety of species, each with high genetic diversity. We may also need to assist the migration of genetic populations and even species into new habitats as they become climatically favourable. Fast-growing plantations of non-native species might grow and fix carbon more rapidly than natural forests, and provide some economic and social benefits, but they will not provide critical habitat for rapidly declining biodiversity.

One thing is for certain: we need to dramatically reduce deforestation globally. Higher-income countries with high carbon footprints should continue to encourage and financially support efforts to slow and reverse tropical deforestation. All nations should support their community-driven efforts to restore degraded forest ecosystems and marginal agriculture lands with a diversity of tree species that provide a variety of resources. If deforestation is to be slowed and reforestation is to succeed, trees must be worth more alive than logged to local people.

Sustainable forest management must also be routinely practiced everywhere. Harvesting crop trees after longer rotations will increase rates of carbon sequestration per year. Thinning stands and using the harvested wood in long-lived products will reduce mortality and improve carbon balance. For example, the use of wood to construct buildings with long lifespans helps store carbon, and provides a substitute for building materials like cement with larger carbon footprints. Partial harvesting and rapid reforestation after harvest will accelerate carbon sequestration, and some stands should be left untouched in the hopes they will become the old growth trees of the future. And we should conserve ancient forests that provide habitat for biodiversity and may be irreplaceable in new climates.

We also need to increase tree cover in urban areas, adapting urban infrastructure and environments to climate change and mitigating some greenhouse gas emissions. Urban forests help cool cities, improve air quality and quality of life, and have positive effects on both mental and physical health. Many cities have lost large numbers of urban trees due to introduced invasive insects and diseases, including the emerald ash borer and Dutch elm disease in North America, and, in Europe, ash dieback due to a fungal pathogen. Adapting urban forests to climate change is best done by planting a diversity of species and cultivars.

Trees have long been a symbol of environmental movements, and tree planting is an important tool in the fight against climate change. But planting trees is not enough. Tree planting will not replace the systematic societal changes to energy, transportation and food production systems needed to slow the pace of climate change and other human impacts on forests.[12] We have many opportunities to help the survival of forests and the species they house, while also mitigating climate change. If we reduce greenhouse gas emissions, reverse deforestation and manage forests sustainably, trees will continue storing carbon cheaply and efficiently, providing habitat for biodiversity and a multitude of products that support human well-being across the globe.

Endnotes

1. C. D. Allen, D. D. Breshears and N. G. McDowell, 'On underestimation of global vulnerability to tree mortality and forest die-off from hotter drought in the Anthropocene', *Ecosphere*, 2015, 6, 129, https://doi.org/10.1890/ES15-00203.1

2. M. L. Goulden and R. C. Bale, 'California forest die-off linked to multi-year deep soil drying in 2012–2015 drought', *Nature Geoscience*, 2019, 12, 632–37, https://doi.org/10.7280/D1DH3B; US Forest Service, *USDA Forest Service Tree Mortality Aerial Detection Survey Results*, Pacific Southwest Region, 2018, https://www.fs.usda.gov/Internet/FSE_DOCUMENTS/fseprd609295.pdf

3. K. Connolly, 'Part of "German soul" under threat as forests die', *The Guardian*, 20 October 2019, https://www.theguardian.com/environment/2019/aug/07/part-of-german-soul-under-threat-as-forests-die

4. Food and Agriculture Organization of the United Nations, *FAO 2012 State of the World's Forests*, Rome: Food and Agriculture Organization, 2012, http://www.fao.org/3/i3010e/i3010e00.htm

5. C. Le Quéré et al., 'Global Carbon Budget 2017', *Earth Syst. Sci. Data*, 2018, 10, 405–48, https://doi.org/10.5194/essd-10-405-2018

6. K. V. Rosenberg, A. M. Dokter, P. J. Blancher, J. R. Sauer, A. C. Smith, P. A. Smith, J. C. Stanton, A. Panjabi, L. Helft, M. Parr and P. P. Marra, 'Decline of the North American avifauna', *Science*, 2019, 366, 120–24, https://doi.org/10.1126/science.aaw1313

7. See 'Everyday Biodiversity' by Jeffrey R. Smith and Gretchen C. Daily in this volume.

8. On air pollution, see 'Air' by Jon Abbatt in this volume. On pollution viewed through an economic lens, see 'Environmental Economics' by Don Fullerton in this volume.

9. IPBES, *Summary for Policymakers of the Global Assessment Report on Biodiversity and Ecosystem Services of the Intergovernmental Science-Policy Platform on Biodiversity and Ecosystem Services*, Bonn, Germany: IPEBS Secretariat, 2019, https://ipbes.net/global-assessment-report-biodiversity-ecosystem-services

10. J. F. Bastin, Y. Finegold, C. Garcia, D. Mollicone, M. Rezende, D. Routh, C. M. Zohner and T. W. Crowther, 'The global tree restoration potential', *Science*, 2019, 365, 76–79, https://doi.org/10.1126/science.aax0848

11. S. L. Lewis, E. T. A. Mitchard, C. Prentice, M. Maslin and B. Poulter, 'Comment on "The global tree restoration potential"', *Science*, 2019, 366, 1–3, https://doi.org/10.1126/science.aaz0388; R. Chasdon and P. Brancalion, 'Restoring forests as a means to many ends', *Science*, 2019, 365, 6448–49, https://doi.org/10.1126/science.aax9539

12. IPCC, 'Summary for policymakers', in *Global Warming of 1.5°C: An IPCC Special Report on the Impacts of Global Warming of 1.5°C Above Pre-Industrial Levels and Related Global Greenhouse Gas Emission Pathways, in the Context of Strengthening the Global Response to the Threat of Climate Change, Sustainable Development, and Efforts to Eradicate Poverty*, ed. V. Masson-Delmotte et al., Geneva: World Meteorological Organization, 2018, https://www.ipcc.ch/sr15/chapter/spm/

Environmental Economics

—

Don Fullerton

The term 'environmental economics' may sound like an oxymoron to those who believe that saving the environment must be based on a moral imperative that ignores financial costs. Yet, when viewed through an economic lens, pollution is ultimately a market failure that can be corrected. Economics can help achieve the *most* environmental protection for any particular amount that society is willing to spend. By identifying market failures that create pollution and helping to design policy proposals that maximize cost-effectiveness, economics can be a powerful tool for environmental protection.

Prior to the first Earth Day in 1970, mainstream economics had well-defined disciplines studying labor markets, international trade and public sector finance (i.e. government tax and spending policy). In contrast, the field of 'environmental economics' did not yet exist, *per se*, although individual economists had certainly explored pollution issues. An early pioneer, Arthur Pigou, pointed out in 1920 that government could impose a tax per unit of pollution (and to this day, economists still refer to a Pigouvian tax on pollution).[1] But Pigou's idea was subsequently challenged by Ronald Coase in 1960, who argued that private interactions could solve pollution problems when property rights are well defined and transactions costs are low.[2] In Coase's scenario, no government regulation meant that polluters could pollute, but victims downstream could simply pay the polluter to cut back emissions to a mutually agreeable level. If, instead, nobody had the right to pollute, then a polluter could pay the victims not to complain — a perfect market! Most often, however, the

https://doi.org/10.11647/OBP.0193.10

reality is not that simple. With many victims downstream, large numbers could 'free-ride' the system, claiming that they don't care about pollution and thus declining to contribute toward the costs of pollution reduction. This outcome is a classic market failure, which can be fixed by government intervention.

Despite these early debates around pollution pricing and other controls, most economists before 1968 largely ignored the study of the environment. But then a remarkable flurry of intellectual activity occurred in the brief period from 1968 to 1974 — the dawning of 'environmental economics'. Many crucial ideas converged within those few years, ending with the 1974 founding of the *Journal of Environmental Economics and Management*. Perhaps ironically, it was a non-economist who wrote the most-cited paper in what was to become environmental economics. In 1968, a biologist named Garrett Hardin published an article in *Science* called 'The tragedy of the commons'.[3] Hardin argued that unregulated use of a commons — a place that everybody can use, with no real owner — could lead to unsustainable exploitation and environmental degradation of the oceans, land and atmosphere. With a growing human population and free access to fishing grounds, for example, each boat takes as many fish as possible — before the others get to it. The ensuing 'tragedy' is the annihilation of the fish stock, or slaughter of the buffalo, or deforestation on a grand scale, or the extinction of many species.

While Hardin's arguments were largely based on biology and demography, to many economists the fundamental problem was a lack of ownership. If some people simply took possession of the resource, then owners could protect their property in the ocean or on land. Under this view, pharmaceutical companies could acquire vast portions of the Brazilian Amazon to protect the rainforest, ensuring preservation of the rich biodiversity necessary to discover and develop valuable life-saving drugs. Such a scheme could work under some circumstances, but history has shown the limits of this approach. Economists had already explained why private markets fail to provide 'public goods' such as roads, law and order, or military defense. A lighthouse is the quintessential example, with two key attributes. First, once the lighthouse is built, its light can provide navigational benefits to many boats in the area who use the resource without ever depleting it — the light is available to additional boats at the same time and at no additional cost. Second, no business could recover the cost of building the lighthouse, because boaters would realize they can see the light whether they pay or not. These free riders cause the private market to fail, even though the social benefits may greatly exceed the costs of building the lighthouse.

In 1968, Hardin didn't use economic terminology, but his reasoning was impeccable: a clean environment represents a public good that provides health and aesthetic benefits to millions of people simultaneously. Once provided, clean air is available to others to breathe at the same time and at no additional cost. Moreover, consumers will not buy clean air, because they can breathe whether they pay or not. With this free-riding behavior, no business would voluntarily pay the costs associated with cleaning up the air. Other firms who do not clean up will be able to charge a lower price for their goods or services, thus gaining a market advantage. Once again, we see the failure of a private market, even though the social benefits of clean air greatly exceed the costs.

In the absence of viable private markets, government can increase social welfare by providing a clean environment; it can regulate firms, require scrubbers, tax pollution and prohibit improper disposal of waste. These clean-up activities certainly have costs, especially for generation of electricity or transportation of goods, and industries may have to cover their costs by increasing product prices. But, *if environmental protection is done wisely*, then collective health benefits can greatly exceed the additional costs to businesses and consumers.

As a thought experiment, consider a particular environmental protection proposal where total health and aesthetic benefits exceed total costs. Suppose also that the benefits and costs are distributed equally across all voters. In this scenario, the proposal would provide a net benefit to everyone, and support for the proposal should be unanimous. Most often, however, the benefits and costs of environmental protection are not shared equally. And therein lies one of the major economic problems of enacting environmental protection. Even for policies with positive net benefits overall, some segments of society receive disproportionate benefits, while others bear disproportionate costs. Critically, economic analysis can be used to measure the distribution of these gains and losses resulting from any proposed policy. It can also help design a policy package that simultaneously achieves pollution reduction *and* desired objectives regarding the distribution of gains and losses.

When it was first established in 1970, the US Environmental Protection Agency (EPA) focused on technological and legal frameworks to control pollution, with little consideration of economic implications. Engineers were employed to determine the 'best' ways to cut pollution, and lawyers wrote regulations requiring the adoption of those recommended technologies. Under this approach, as it developed in the immediate aftermath of the first Earth Day, environmental protection was viewed as a moral imperative, and costs were not taken into account. In contrast, the early pioneers in environmental economics devoted significant attention to analyzing both the costs and the benefits of different environmental protection schemes. They often found that costs of actual legislative and regulatory changes were more than three times as high as those for alternative policies that would achieve the same degree of environmental protection. In other words, more economically efficient approaches could lead to greater environmental protection for the same level of financial investment.

Enter the ideas of John Harkness Dales. In 1968, Dales published a brilliant idea for minimizing the cost of achieving any given degree of environmental protection.[4] Government could limit the total amount of pollution at an appropriate low level, print a fixed number of permits or licenses, and let polluters bid for the permits or trade with each other. The key innovation of Dales' idea was to recognize that a particular required mitigation technology cannot logically be 'best' in all different circumstances. Policymakers in the nation's capital cannot possibly know as much about production technologies as the engineers inside each firm, especially when those technologies vary across firms. The same pollution reduction could be achieved by letting each firm determine their own 'best' method.

As an example, regulators might require the most advanced (and likely most expensive) flue-gas scrubber to remove sulfur dioxide from emissions of coal-fired electricity generating plants, but cost-minimizing engineers within the firm might be able to cut pollution the same amount at lower cost. They could switch from high-sulfur coal to low-sulfur coal, or from coal to natural gas, or change the dispatch order between coal plants and gas plants, or use renewable power like wind and solar. If the goal is a target pollution reduction, then the method of reduction should not matter. Moreover, not all of those strategies need to be available to every firm. With permit *trading*, a firm with limited options can essentially pay a different firm to do their required pollution reduction, through a so-called 'cap-and-trade' approach. Imagine ten firms that each hold 1,000 one-metric-ton permits for sulfur dioxide emissions. Together, these firms are collectively limited to 10,000 metric tons of emissions, but they do not all have to cut by the same amount. A firm with only fossil-fuel-fired power plants could switch some output from coal to gas plants, while also buying additional permits from some other firm in sunny Arizona with abundant solar power.

With a single market price, say $100 per metric ton of sulfur dioxide emissions, a tradeable permit system provides incentive for any firm to develop emission reduction strategies that cost less than $100 per metric ton. Firms would bypass any technology costing more than $100 to reduce emissions per metric ton, choosing the more economical approach of simply buying an emissions permit instead. The exact same argument can be applied to an emission tax of $100 per ton. Both emissions taxes and permits represent pollution-pricing policy; in either case, only the cheapest pollution abatement methods are chosen, minimizing the total cost of achieving any given target pollution reduction. If a pricing policy could effectively reduce pollution by the same amount at lower cost, it would allow policymakers to choose a more ambitious target for the same overall expenditure. In other words, the same dollar cost could be used more efficiently to achieve greater pollution reduction.

The flurry of important new ideas in the emerging field of environmental economics continued through the early 1970s. In 1971, William Baumol and Wallace Oates described various approaches to implementing pollution pricing policies.[5] And in 1972, David Montgomery showed exactly what conditions would be necessary for a permit policy to minimize the total social cost of pollution abatement.[6] These contributions culminated in the significant 1974 paper by Martin Weitzman that explored the difference between taxation and permit policy as a means of pollution pricing.[7] A tax on pollution fixes the price of pollution, but it does not necessarily limit the total quantity of emissions. Firms facing a fixed price will decide their quantity of pollution and thus the total amount they are willing to pay for it. If policymakers knew the total quantity of pollution that would result under a given taxation scheme, they could fix that quantity of pollution by printing a fixed number of tradable permits. Under this permit system, the market would be expected to produce an equivalent pollution price for the same quantity of emissions. But the critical difference, pointed out by Weitzman, relates to future *uncertainty* as market conditions evolve. Limiting the quantity of pollution through a permitting system is great for ensuring a clean environment, but firms cannot be sure what price they will have to pay in the future. That uncertainty can inhibit investment and reduce growth, raising costs. On the other hand, setting the price of pollution through taxation is great for ensuring a known cost of production (and thus certainty for investors), but this approach creates uncertainty about the total amount of resulting pollution.

Which policy, taxation or permitting, better maximizes total social welfare — accounting for all economic and environmental costs and benefits? The answer depends on the relative impacts of uncertain economic costs as compared to uncertain environmental costs. We face many environmental problems ranging from contaminated water, climate change, endangered species and local air pollution. At one end of the spectrum, where the quantity of pollution is not critical, pollution that reduces aesthetic amenities (like visibility) might best be handled by a tax that fixes the price of pollution and avoids the risk of very high costs on business and consumers. At the other end of the spectrum, some types of pollution have critical thresholds, like the 1952 Great Smog of London that caused thousands of deaths.[8] These pollutants might best be handled by a permit system that fixes the quantity of emissions *below* that critical threshold (even though the price per metric ton could end up quite high).

In the decades since the first environmental economics journal started in 1974, thousands of scholarly articles have been published examining multiple aspects of this field. New theoretical ideas have added to those described above, and the discipline has become more empirically driven by advances in 'big data'. For example, observations from satellite remote sensing have been used to estimate the deforestation effects of various land-use changes, from agricultural policy and mining to various attempts at reforestation. Large data sets have also been important to quantify the effects of environmental policy on industrial output, productivity, employment and growth, in an effort to maximize the cost-effectiveness of various environmental policy approaches.

Over the past half-century, environmental economics has earned its keep. Academic ideas like permit trading have been put to the test in many practical applications, starting in the US with the 1990 Clean Air Act Amendments that initiated sulfur dioxide permit trading and largely eliminated the acid rain problem. Permit policies for carbon dioxide emissions began in 2005 for the European Union, and in 2006 for California. Carbon taxes have now been enacted in a dozen countries and in three Canadian provinces. These policies are currently generating much data for further analysis by environmental economists trying to help design better polices that can help protect Earth's environment at lower cost. These economic approaches can play a huge role in aiding our transition to more sustainable societies.

Endnotes

1. A. C. Pigou, *The Economics of Welfare*, London: Macmillan & Co., 1920.

2. R. H. Coase, 'The problem of social cost', *Journal of Law and Economics*, 1960, 3, 1–44, https://doi.org/10.1086/466560

3. G. Hardin, 'The tragedy of the commons', *Science*, 1968, 162, 1243–48, https://doi.org/10.1126/science.162.3859.1243

4. J. H. Dales, *Pollution, Property, and Prices: An Essay in Policy-Making and Economics*, Toronto: Toronto University Press, 1968.

5. W. J. Baumol and W. E. Oates, 'The use of standards and prices for the protection of the environment', *Swedish Journal of Economics*, 1971, 73, 42–54, https://doi.org/10.1007/978-1-349-01379-1_4

6. W. D. Montgomery, 'Markets in licenses and efficient pollution control programs', *Journal of Economic Theory*, 1972, 5, 395–418, https://doi.org/10.1016/0022-0531(72)90049-x

7. M. L. Weitzman, 'Prices vs. quantities', *Review of Economic Studies*, 1974, 41, 477–91, https://doi.org/10.2307/2296698

8. On the Great Smog of London, see also 'Air' by Jon Abbatt in this volume.

Air

—

Jon Abbatt

Life as we know it is possible thanks to the thin film of gases in Earth's atmosphere, which distinguishes our planet from all others in the solar system. A ride on an airplane offers the opportunity to ponder this remarkable atmospheric eggshell around us. Only a few minutes after takeoff, we reach cruising altitude around 10,000 m above sea level, with two-thirds of the atmosphere below us. This lowest portion of the atmosphere — the troposphere — contains gases that support life on our planet, as well as all the pollutants that damage our lungs. The troposphere is also home to the greenhouse gases that warm us and the clouds that cool us. If our plane were to fly another 10 or 20 km higher, it would pass through the ozone layer in the stratosphere, which filters out biologically damaging ultraviolet radiation from the Sun. As we look back to the first Earth Day in 1970, we can ask ourselves how Earth's atmosphere has changed, and what the future may yet hold.

For much of human history, our Earth-bound species has largely taken the air around us for granted, the atmosphere conceived of as an invisible and infinite conduit to the heavens above. This notion was radically challenged with the rise of industrialization, when coal-darkened skies became common in cities across North America and Europe. The 1952 acidic fog episode in London provides a famous example. At that time, England was burning poor quality bituminous coal, creating high levels of carcinogenic soot particles in the air, as well as sulfur dioxide, which is harmful to our respiratory system. In early December of 1952, low winds created a stagnant pool of air that trapped the coal fumes

     https://doi.org/10.11647/OBP.0193.11

over London for four consecutive days. Conditions grew so bad that people could not see across the road, and outdoor activities were restricted. Several thousand people died as a direct effect of the air pollution (primarily those with pre-existing respiratory problems), and many thousands more suffered adverse longer-term health effects. Response to the Great Smog of London was swift. The UK passed its Clean Air Act only four years later, in 1956, which mandated the burning of cleaner fuels and led to a rapid improvement in air quality in cities across the country.

Episodes like the Great Smog of London, along with rising public awareness of environmental pollution, led to the first Earth Day in 1970. In the subsequent fifty years, environmental science and public policy have converged to address a variety of atmospheric pollution phenomena.[1] This is perhaps best illustrated by international efforts to save the ozone layer. In the early 1970s, only a handful of scientists cared about ozone. This molecule (three oxygen atoms bonded together) was known to block harmful ultraviolet light from the Sun, but there was no indication that human activity could affect its abundance. After all, ozone mostly existed in the stratosphere, 20 to 30 km above Earth's surface. At the time, it was not known that industrial chemicals, known as chlorofluorocarbons (CFCs), were percolating upwards towards the stratosphere. These compounds, invented in the 1950s, were initially viewed as a shining example of human industrial ingenuity — non-toxic, non-flammable substances with many uses in cleaning, refrigeration and aerosol sprays. Yet, in the decades that followed, the true environmental impact of these compounds would come to capture global attention.

The first measurements of global CFC abundance were reported in the 1970s by James Lovelock, who would later propose the Gaia hypothesis of Earth as a self-regulating environment sustaining life. Even though the CFC sources were largely in the northern hemisphere, where populations and industry are dominant, the abundance of these compounds was just as high south of the equator. This was one of the first indications that anthropogenic pollutants experience widespread global transport across geopolitical boundaries. Attempting to explain the behavior of CFCs in the atmosphere, researchers discovered that they decompose in the stratosphere, releasing chlorine that catalyzes ozone destruction.[2] At the time, CFCs were being used around the world in a wide range of applications, and their atmospheric concentrations were rising dramatically. Once released, there was no easy way to remove these molecules from the atmosphere.

The problem of CFC-driven ozone destruction captured public attention in the mid-1980s, with the discovery of a massive ozone hole over the Antarctic continent. This phenomenon, seen as a large and recurrent loss of ozone in the region each spring, is driven by the chlorine released from CFCs. Its location over the southern pole is attributable to unique meteorological factors that isolate cold air masses over Antarctica and make them particularly susceptible to chlorine-mediated ozone loss. Ground-based observations of the ozone layer over the Antarctic continent, which started in the late 1950s as part of the International Geophysical Year, were the first to detect the ozone hole in the mid-1980s.[3] Subsequently, satellite-derived images of the widespread Antarctic ozone hole became emblematic of human impacts on the environment — a dystopian view of human technology gone horribly wrong.

As terrifying as detection of the ozone hole was, global action was swift and extremely effective. The Montreal Protocol, first signed in 1987 and amended a number of times thereafter, led to the banning of CFC production globally.[4] Other ozone-depleting substances, such as methyl bromide, once used to fumigate strawberry fields, have also since been banned. Ozone-friendly CFC replacement compounds are now widely used, with society barely noticing the transition. Yet, the lifetime of CFCs is so long — fifty to one hundred years — that significant ozone depletion continues over the Antarctic and more slowly at mid-latitudes.[5] Once the CFCs have been naturally cleansed from the atmosphere, ozone levels will hopefully return to those present on the first Earth Day. The enactment of the Montreal Protocol and the saving of the ozone layer has undoubtedly adverted millions of cases of skin cancer. An environmental success story indeed!

Throughout the 1970s, as stratospheric ozone loss became a cause for significant concern, ozone began to accumulate in the lowest layers of the atmosphere. Increasing ground-level ozone concentrations, first identified in Los Angeles and subsequently in other large cities around the world, resulted from chemical reactions between organic molecules (including gasoline fumes) and nitrogen oxides (emitted by car engines) in sunlight. The resulting ozone caused significant damage to a variety of organic materials, from rubber tires and windshield wipers to people's breathing passages. Los Angeles, home to plenty of sunlight and automobiles, became the posterchild of photochemical air pollution, or 'smog', as it came to be known.

In response to the smog crisis of the 1970s and 1980s, California developed air pollution control strategies that are now widely adopted across the globe. Catalytic converters were added to automobile exhaust systems to remove organic and nitrogen oxide vapours, and internal combustion engines were designed to use gasoline that combusts much more efficiently, with computer-controlled tuning of air-to-fuel ratios. These measures have had a dramatic effect on air quality in Los Angeles and other major cities. Whereas the Los Angeles' automobile population has grown enormously from the 1970s, when the smog pollution was at its worst, ground-level ozone concentrations have dropped by roughly two thirds over the last forty years. In the 1970s, visitors to Los Angeles were surprised to find that the city is ringed by a range of mountains, which were infrequently invisible through the haze. Today, visibility is much improved. Additional progress will be made as gasoline-powered engines give way to electric and hydrogen-based vehicle propulsion systems. Even so, ozone production will continue from organic precursor molecules derived from a variety of consumer products, such as paints, solvents, personal care products and indoor cleaning agents.[6] These sources, long overlooked, are currently unregulated, posing an on-going challenge for long-term air quality improvement. But it seems only a matter of time before these chemicals, like automobile exhaust and CFCs, will also be subject to strict environmental regulation.

The factors that led to urban smog in cities around the world also created additional atmospheric pollution problems. In the 1970s, forests and lakes were dying in northeastern North American and northern Europe as a result of acid rain produced from sulfur and nitrogen oxides released from coal burning and vehicle exhaust. In extreme cases, the acidity of rainwater approached that of vinegar, and this low pH precipitation was deposited onto land and water surfaces with devastating effects. The strong acidity had direct biological impacts on marine and terrestrial ecosystems, and a variety of indirect effects, including the leaching of toxic metals from soils. There was also another, very visible, manifestation of acid rain in cities around the world, as low pH rain dissolved the ornate limestone structures on historical buildings and attacked the steel beams of bridges. Motivated by the mass protests on Earth Day 1970, new environmental regulations mandated the widespread use of smokestack scrubbers and cleaner coal, both of which led to a significant decrease in sulfur dioxide emissions and the associated acid rain.

Similarly effective action was taken to combat atmospheric lead pollution. The environmental toxicity of lead, from cookware to paint, has impacted human societies for millennia, and lead poisoning was famously suggested as a cause for the decline of the Roman Empire. But it was not until the twentieth century, when lead became widely used as a gasoline additive that the concentrations of this metal began to increase on a global scale through long-range atmospheric transport. As with CFCs and acid rain, the solution to atmospheric lead contamination was clear and remarkably effective. In a 1996 amendment to the US Clean Air Act, lead was banned from all gasoline products, and over the next two decades, human blood levels of lead dropped by more than 80%.

Unlike CFCs and lead, air-borne particulate matter continues to be an important component of air pollution.[7] These small solid and liquid particles, much smaller than the width of a human hair, have serious health consequences when present in high abundance. This is particularly true for the smallest size class of particles, which are readily inhaled into the lungs. The landmark 'Harvard Six Cities' study, initiated in the 1970s, has continually monitored the mortality of people living in six American cities.[8] After correcting for occupational hazards and smoking rates, the study has shown a strong correlation between rates of excess mortality and high amounts of air-borne particulate matter. This result has been confirmed in numerous cities, and it is now clear that atmospheric particles are one of the leading causes of shortened lifespan worldwide, leading to millions of excess deaths per year. It remains to be determined which chemical compounds of the thousands present in atmospheric particulate matter are leading to these negative health outcomes.

Particles are emitted into the atmosphere from both natural and industrial sources, both of which are likely to continue increasing for the foreseeable future. Natural sources of atmospheric particulate matter include sea spray, desert dust and wildfires. The frequency and intensity of wild-fires is predicted to increase under a warming climate, as is the expansion of some global deserts. Both of these processes should act to increase sources of atmospheric particles. Even in the absence of fires, forests can be a source of atmospheric particles by emitting gaseous organic molecules that undergo chemical transformations in sunlight. The Great Smoky Mountains in Tennessee and North Carolina get their name from this phenomenon.

Over the past five decades, human-derived sources of atmospheric particulates have increased dramatically. These anthropogenic particles are derived from both fossil fuel and vegetation burning, as well as specific industrial activities such as metal smelting. The world's population is more urbanized than it has ever been, with over half of us now living in cities. The growth of megacities with populations of more than ten million people has been remarkable, with most of these cities in industrially developing countries. These urban centers have extremely high air particle levels, resulting from the burning of dirty coal and agricultural wastes, widespread street cooking and the unregulated use of many commercial products, including small motorcycles without air pollution controls. Moreover, indoor air quality remains a serious problem in millions of homes around the world in which cooking is still performed over inefficient stoves using wood, coal or dung fuels. As reported by the recent Global Burden of Disease study, the air quality in or near these homes is one of the leading causes of pollution-related death globally, particularly for women and children.[9] The implementation of better ventilation, more efficient cook stoves, and cleaner fuels are needed to address this global health problem.

The examples of London and Los Angeles illustrate how we have dealt with air pollution crises in the past. Technological solutions exist, and with increasing wealth around the world, a growing middle-class will demand cleaner air as a fundamental human right. Indeed, recent widespread protests in China over poor urban air quality garnered significant international attention, prompting the government to pledge new environmental protection measures. When and how governments around the world deliver on this promise remains uncertain, but it would seem to be only a matter of time before air pollution levels in the world's new megacities are reduced.

While ozone depletion, acid rain and urban air pollution are being addressed, enhanced global warming associated with atmospheric release of greenhouse gases — most notably carbon dioxide, methane and nitrous oxide — remains a larger, daunting challenge. Ice core records show that the concentrations of these gases in the atmosphere are significantly higher than at any time over the last 800,000 years, with a rate of increase that may be unprecedented in Earth's history.[10] Unlike CFCs or lead, whose industrial sources could be traced to specific sources (spray cans and leaded gasoline, for example), CO_2 emissions result from the combustion of all fossil fuels, from coal and oil to wood and natural gas. For this reason, a reduction in CO_2 emission requires nothing less than a whole-scale transformation of global energy production systems.

The global warming challenge mirrors previous global air pollution threats. In the case of ozone depletion, lead, acid rain and smog, society recognized the central role played by key compounds — CFCs, sulfur dioxide, nitrogen oxides and particulate matter — and policies were put in place to successfully control these emissions. We can only hope that these previously successful approaches can provide a template for tackling global warming and transforming our energy supply network, with sound science and technological innovation tied to effective public policy. Though it has been suggested that carbon capture may be necessary to limit global warming to 1.5°C, we hopefully will not need to rely on other geoengineering schemes — such as injection of aerosol particles into the stratosphere to block incoming sunlight — to avert the most dire warming scenarios. Indeed, it is heartening to see the price for wind and solar energy rapidly dropping, to the point that these non-carbon energy sources are now economically competitive with fossil fuel energy in many places. If we apply the same focus and energy used to address air pollution issues over the past half-century, we can remain optimistic that the one-hundredth anniversary of Earth Day may see the atmosphere returning towards its pre-industrial character.

Endnotes

1. D. Davis, M. Bell and T. Fletcher, 'A look back at the London smog of 1952 and the half-century since', *Environmental Health Perspectives*, 2002, 110, A734–A735, https://doi.org/10.1289/ehp.110-a734

2. M. J. Molina and F. S. Rowland, 'Stratospheric sink for chlorofluoromethanes — chlorine atomic-catalyzed destruction of ozone', *Nature*, 1974, 249, 810–12, https://doi.org/10.1038/249810a0

3. J. C. Farman, B. G. Gardiner and J. D. Shanklin, 'Large losses of total ozone in Antarctica reveal seasonal ClOx/NOx', *Nature*, 1985, 315, 207–10, https://doi.org/10.1038/315207a0

4. Available at https://ozone.unep.org/treaties/montreal-protocol-substances-deplete-ozone-layer/text

5. World Meteorological Organization, *Scientific Assessment of Ozone Depletion: 2018*, Global Ozone Research and Monitoring Project, Report 58, Geneva: World Meteorological Organization, 2018, https://www.esrl.noaa.gov/csd/assessments/ozone/2018/downloads/2018OzoneAssessment.pdf

6. B. C. McDonald et al., 'Volatile chemical products emerging as largest petrochemical source of urban organic emissions', *Science*, 2018, 359, 760–64, https://doi.org/10.1126/science.aaq0524

7. U. Poschl, 'Atmospheric aerosols: Composition, transformation, climate and health effects', *Angewandte Chemie-International Edition*, 2005, 44, 7520–40, https://doi.org/10.1002/anie.200501122

8. D. W. Dockery, C. A. Pope, X. P. Xu, J. D. Spengler, J. H. Ware, M. E. Fay, B. G. Ferris and F. E. Speizer, 'An association between air pollution and mortality in 6 United States cities', *New England Journal of Medicine*, 1993, 329, 1753–59, https://doi.org/10.1056/NEJM199312093292401

9. S. S. Lim, T. Vos, A. D. Flaxman et al., 'A comparative risk assessment of burden of disease and injury attributable to 67 risk factors and risk factor clusters in 21 regions, 1990–2010: a systematic analysis for the Global Burden of Disease Study 2010', *The Lancet*, 2012, 380, 2224–60, https://doi.org/10.1016/S0140-6736(12)61766-8

10. IPCC, 'Climate change 2013: The physical science basis', in *Contribution of Working Group 1 to the Fifth Assessment Report of the Intergovernmental Panel on Climate Change*, ed. T. F. Stocker, D. Qin, G.-K. Plattner, M. Tignor, S. K. Allen, J. Boschung, A. Nauels, Y. Xia, V. Bex and P. M. Midgley, Cambridge, UK: Cambridge University Press, 2013, 383–464, https://www.ipcc.ch/site/assets/uploads/2018/02/WG1AR5_all_final.pdf

Geoengineering

—

Douglas G. MacMartin and Katharine L. Ricke

When people think about responding to climate change, they typically think about reducing emissions of carbon dioxide (CO_2) and other heat-trapping greenhouse gases. Had we started on a path to reducing these emissions at the time of the first Earth Day — when the science was already indicating that our emissions would cause global warming — then climate change might be behind us today. Instead, fifty years later, our collective emissions are higher than they have ever been. Cutting emissions is absolutely essential, but it is no longer sufficient.[1] We must transform our entire global energy infrastructure, not just to reduce our emissions of greenhouse gases, but to eliminate them altogether. That won't happen overnight, and even if we succeed in that challenge over the next few decades (which we must), there will still be substantial global warming. This is our new reality in 2020.

Because CO_2 remains in the atmosphere for a long time, reaching zero emissions won't eliminate climate change, it will just stop making the problem worse.[2] Like a driver careening towards the car in front of us, the first thing we must do is to take our foot off the gas pedal. But that alone won't necessarily prevent the damage. The next step is to apply the brakes — and quickly — to lessen the impending impact. And even then, we might need airbags to avoid the worst possible consequences.

Over the past several decades, as our failure to limit greenhouse gas emissions has become ever more apparent, there has been increasing interest in applying the brakes on

    https://doi.org/10.11647/OBP.0193.12

global warming by removing CO_2 from the atmosphere after it has been emitted. This set of ideas, known as carbon dioxide removal (CDR) or negative emissions technologies,[3] includes 'natural' methods such as planting trees, or changing agricultural practices to store more carbon in the soil; artificially fertilizing the oceans to encourage phytoplankton blooms that consume CO_2 and sequester some of it in the deep ocean; chemically capturing CO_2 from the air through reaction with various minerals; or enhancing the rate of weathering of rocks, the natural process that will ultimately remove atmospheric CO_2 over the coming millennia.[4] The challenge today is that while many CDR approaches have promise, none of them currently satisfies three essential criteria.

First, carbon removal needs to be scalable. Each tree planted, for example, will ultimately absorb something in the order of 1 ton of CO_2 over the next forty years. By comparison, we are currently emitting nearly 1300 metric tons of CO_2 *per second*. There is roughly a trillion more tons of CO_2 in the atmosphere than there was at the dawn of the industrial revolution, and, if we ramp down to zero emissions over the next twenty-five years, we will have emitted half that amount again. There simply isn't enough available land for tree planting alone to solve the problem we've created.[5] There are similar scaling limitations on other carbon removal approaches as well, in particular those that most closely mimic natural ecological processes.

Second, carbon removal needs to be reasonably economical. While planting trees might be relatively cheap, the current projected costs for more globally scalable approaches — such as direct capture of CO_2 from the air — are $100 or more per ton. At this price, removing just one year's worth of our current emissions would cost $4 trillion, about 20% of the United States GDP.

And third, carbon removal should not create local impacts that are potentially worse than climate change itself. The generation of bio-energy from plants would remove carbon from the atmosphere if the resulting CO_2 were captured from the flue gas and stored underground. But deploying this approach at the scale required to have a global impact would require either a large-scale transformation of natural ecosystems, or a massive diversion of land towards energy crops, resulting in competition for both food and water. In the case of ocean fertilization, the additional carbon transported into the deep ocean would stimulate oxygen consumption that could render some regions inhospitable to animal life.[6] Clearly, such unintended consequences must be factored into any future considerations.

With further research and development, there are carbon removal approaches that, when implemented together, might avoid all three of the challenges above. But at the same time, it would be foolhardy to assume that these approaches to CO_2 reduction can be relied on with certainty to avoid future climate change. And it would be even more unwise to continue to emit CO_2 today on the assumption that our children and grandchildren can figure out how to remove it.

We are thus left with no certain pathway to avoid serious climate change impacts. The most optimistic scenarios include both a rapid transformation of our entire global energy and agricultural systems, and a massive scale-up of 'negative' emissions using CDR technologies that currently do not exist. This is clearly a daunting task, both technically and politically, yet this is required if we are to have even reasonable odds of avoiding significant warming. The challenge is compounded by the fact that we don't know precisely how much the climate will continue to warm, or how bad the impacts of that warming will be. Most people carry fire insurance on their house, despite the odds of a fire being less than 1%. Yet, even the optimistic scenarios do not ensure that we can meet temperature targets and avoid the worst potential impacts of predicted climate change. We are, quite literally, gambling with the future of the planet.

In the face of this future uncertainty, there is another tactic — in addition to mitigation and carbon dioxide removal — that might provide a kind of planetary insurance. This approach, known as 'solar geoengineering', aims to reduce global warming by decreasing the amount of incoming energy from the Sun.[7] These ideas are not new, indeed they were discussed in the mid-1960s when US President Lyndon B. Johnson was briefed on climate change. But solar geoengineering remained mostly on the fringe until 2006, when Paul Crutzen, who was awarded a Nobel Laureate for his work in atmospheric chemistry, suggested that it be taken seriously.[8]

At its most basic level, solar geoengineering seeks to modify the radiation balance of Earth. When left to its own devices, the planet reaches an energy equilibrium state, with the amount of energy received from the sun closely balanced by the amount of energy sent back into space through reflected sunlight and emission of thermal radiation (heat). The reason the climate is warming today is that increased greenhouse gas concentrations in the atmosphere are making it harder for Earth's thermal energy to escape back to space. Since the Earth is now receiving more energy than it is emitting, it must warm up (increasing thermal losses) until the input and output are back in balance. Reducing atmospheric greenhouse gas concentrations deals with the imbalance directly, but reducing the amount of incoming energy could address the other side of the balance sheet. If we could deliberately reflect roughly 1% of the sunlight currently hitting Earth's surface back to space before it is absorbed, we would cool the planet enough to counteract all the warming from our past greenhouse gas emissions.

Just how difficult would it be to accomplish this? While 1% doesn't sound like a lot, consider, for perspective, that the entire continental US covers about 2% of Earth's surface. So, we cannot achieve this additional reflection by doing things like painting roofs white; there just aren't enough roofs. There are, however, at least two proposed approaches that could plausibly reflect enough sunlight to significantly influence global climate.

One such approach would mimic the cooling effect that occurs after large volcanic eruptions, such as the eruption of Mount Pinatubo in the Philippines. On June 15 of 1991, an explosive eruption from Pinatubo emitted large amounts of sulfur dioxide high into the atmosphere, where the gas underwent chemical reactions to produce reflective sulfate aerosols (small droplets or particles). If the gas had been released into the troposphere (the lower atmosphere), the resulting aerosols would have been rained out within weeks, with relatively little cooling effect. But higher up in the stratosphere — around 20 km above Earth's surface — the air is stable and dry, and the aerosols can persist for a year or more. These stratospheric aerosols, which were clearly visible in satellite imagery, reflected enough solar radiation back into space to decrease global temperatures by 0.3–0.5°C over the following year.

It is, in principle, possible to deliberately mimic this process of solar reflectance (without all of the ash and other negative impacts of a volcanic eruption). The stratospheric-aerosol approach would cool the planet, and would thus counteract many — but not all — of the impacts of climate change. We don't currently have aircraft that fly high enough with the capacity to deliver a useful payload, but these engineering challenges appear surmountable. In fact, one of the concerns with this idea is that the direct costs might be low enough to make the idea more enticing than it should be!

Another solar geoengineering idea is to enhance the formation of reflective low clouds over the ocean. Satellite imagery reveals that ships in some parts of the ocean leave behind 'cloud tracks' that can persist for up to a week. This phenomenon occurs when aerosol pollution from the ship enhances the formation of cloud droplets, either creating a cloud where none previously existed, or making more, smaller droplets that make existing clouds 'brighter'. In either case, the result is the same; more sunlight is reflected back to space. Achieving this effect does not necessarily require adding pollution; spraying salt water into the right type of clouds might be sufficient.

Spraying salt water into clouds may be more benign than adding sulfate to the stratosphere, but we don't understand the physics of cloud-aerosol interactions well enough to know how well this approach might work. Cloud brightening also comes with its own set of issues. While stratospheric aerosols spread roughly uniformly across the globe, marine clouds that can be brightened might only exist over about 10% of the Earth's surface. Achieving the same global cooling effect through cloud enhancement would require much larger changes over smaller areas, resulting in potentially significant impacts on regional weather patterns.

Beyond any technical challenges of solar geoengineering, there are other significant questions to be addressed, from the details of its physical impacts, to broader societal issues such as public acceptability, ethics and international relations. For example, both cloud-brightening and the introduction of stratospheric aerosols have the potential to change precipitation patterns. Climate models suggest that these precipitation changes will typically be smaller than those we would experience if we allowed climate change to grow without geoengineering. But that might not be true everywhere, and there is still considerable uncertainty in model predictions. In addition, stratospheric aerosols could delay the recovery of the ozone layer through their interactions with the long-lived chlorine compounds (CFCs) that were phased out by the 1987 Montreal Protocol.[9] And what goes up must come down — so there may be ecological impacts as sulfate aerosols are eventually returned to Earth's surface in the form of acid rain (though the amount of acid rain would likely be a small increment over today's background levels). We simply don't know enough today to adequately inform future decisions. More research might uncover reasons why geoengineering would always be a bad idea, or might conclude that the consequences of not deploying these approaches outweigh these concerns.

More challenging still are the societal and governance questions.[10] If deployed, solar geoengineering would affect everyone on the planet. Who would decide, and how? Whose voices would be heard; whose interests would matter?

If CO_2 emissions continue unabated, an increasing amount of geoengineering will be required to compensate. Future generations would be committed to maintaining the deployment practically indefinitely; if they ever stopped, the climate would rapidly warm to where it would have been without geoengineering. On top of that, some impacts of our anthropogenic emissions wouldn't be addressed at all by solar geoengineering, such as the ocean acidification driven by high atmospheric CO_2 concentrations. Despite these obvious concerns, there will be some who want to use a geoengineering option as a shortsighted excuse not to cut CO_2 emissions. How can we ensure that this approach is considered only as a supplement and not as a substitute? To answer this question, it is essential that scientific research into geoengineering goes hand in hand with the development of international governance capacity to make sound decisions.

Returning to the car-accident analogy, solar geoengineering is akin to air bags. It doesn't quite deal with the underlying problem of an impending impact — in our case, of having added greenhouse gases to the atmosphere. No-one would sit in their car and set off their air bag for fun, and, similarly, it only makes sense to consider the side-effects of solar geoengineering in the context of climate change. But it is possible that geoengineering could reduce some of the worst effects of climate change, and thus mitigate suffering, particularly for the world's most vulnerable inhabitants. For ecosystems without a capacity to adapt to rapidly changing conditions, a climate response plan that includes solar geoengineering may be the only way to avoid extinctions.

It wouldn't make sense to force society to choose between installing air bags in cars and enforcing speed limits. Similarly, we don't have to choose between cutting emissions, developing and deploying methods to remove CO_2 from the atmosphere, and conducting research to better understand solar geoengineering. Indeed, geoengineering approaches only make sense in conjunction with cutting greenhouse gas emissions. Had we been working diligently to reduce our CO_2 emissions over the last fifty years, perhaps we wouldn't need to think today about additional approaches to climate change response. Even if solar geoengineering is eventually deployed to help limit the impacts of climate change, we must strive for a future Earth Day when the excess atmospheric CO_2 will have been removed and solar geoengineering is no longer needed.

Endnotes

1. IPCC, 'Summary for policymakers', in *Global Warming of 1.5°C: An IPCC Special Report on the Impacts of Global Warming of 1.5°C Above Pre-Industrial Levels and Related Global Greenhouse Gas Emission Pathways, in the Context of Strengthening the Global Response to the Threat of Climate Change, Sustainable Development, and Efforts to Eradicate Poverty*, ed. V. Masson-Delmotte et al., Geneva: World Meteorological Organization, 2018, https://www.ipcc.ch/sr15/chapter/spm/

2. See also 'Carbon' by David Archer in this volume.

3. National Academy of Sciences, *Climate Intervention: Carbon Dioxide Removal and Reliable Sequestration (2015)*, Washington, DC: National Academies Press, 2015, https://doi.org/10.17226/18805.

4. See also 'Carbon' by David Archer in this volume.

5. See also 'Forests' by Sally N. Aitken in this volume.

6. See also 'Oceans 2020' by David M. Karl in this volume.

7. National Academy of Sciences, *Climate Intervention: Reflecting Sunlight to Cool Earth*, Washington, DC: National Academies Press, 2015, https://doi.org/10.17226/18988

8. P. J. Crutzen, 'Albedo enhancement by stratospheric sulfur injections: A contribution to resolve a policy dilemma?', *Climatic Change*, 2006, 77, 211–19, https://doi.org/10.1007/s10584-006-9101-y

9.   Available at https://ozone.unep.org/treaties/montreal-protocol-substances-deplete-ozone-layer/ text

10. J. Pasztor, 'The need for governance of climate geoengineering', *Ethics & International Affairs*, 2017, 31, 419–30, https://doi.org/10.1017/S0892679417000405

Ice

—

Julian Dowdeswell

With an average surface temperature of 15°C (and rising), much of our planet is inhospitable to ice. Today, less than 2% of Earth's water exists in a frozen form, locked up in glaciers and ice sheets, sea ice and permafrost. This 'cryosphere' is critically important for controlling global sea level and the distribution of the planet's fresh water, yet it has always existed in a rather perilous state. In contrast, the ice caps on Mars and the frozen surface of Jupiter's moon, Europa, enjoy a much colder and more stable existence. To understand the impacts of climate change on Earth's cryosphere, it is necessary to examine the different components of our icy world separately, for each has its own sensitivity to local and global forces.

Land-based glaciers and ice sheets develop when winter snowfall persists through successive summers, building up and compacting under its own weight into frozen layers that may be hundreds and sometimes thousands of meters thick. Inputs of snow on the ice-sheet surface are balanced by losses in the form of basal melting and iceberg production at ice-sheet marine margins and, in some milder areas, by surface melting and water runoff. Because of their origins from snow, glaciers and ice sheets contain fresh water. In total, about 70% of the planet's fresh water is presently locked away in these glaciers and ice sheets.

     https://doi.org/10.11647/OBP.0193.13

Today, the great ice sheets of Antarctica and Greenland cover areas of 13.7 and 1.7 million km², respectively. The Antarctic ice sheet has an average thickness of about 3 km and a maximum of almost 5 km, with an approximate total ice volume of 30 million km³. On Greenland, ice reaches about 3 km deep and encompasses a total volume of about 3 million km³. Beyond these two great ice sheets, smaller ice caps and glaciers are present on many Arctic and Antarctic islands and in mountain chains around the globe, from the Himalayas and European Alps to the South American Andes. Together, these smaller ice masses cover about 700,000 km² (about 0.1%) of Earth's land surface.

Ice sheets and glaciers are not static entities, frozen in time and place. Rather, they are dynamic structures that move in response to various forces, including their own massive weight. Although ice exists as a solid, it deforms and flows slowly under pressure, similar to a metal that softens as its melting point is approached. In response to gravitational forces, glaciers flow by internal deformation at speeds of just a few meters per year, moving down-slope from higher elevations towards sea level. Glacier flow can be many times faster than this, however. Continuously fast-flowing ice streams within ice sheets, and so-called 'surging glaciers', where fast flow is intermittent, can move at hundreds and sometimes even thousands of meters per year when lubricating water reduces friction at their beds. Where ice sheets reach the sea, large table-like icebergs are broken off from edges of ice sheets. These icebergs, with underwater keels sometimes hundreds of meters deep, can drift for hundreds and sometimes several thousands of kilometers, well beyond the icy coasts where they originated. Icebergs from Antarctica have occasionally been observed off New Zealand's South Island, and Arctic icebergs often travel south into the North Atlantic, creating hazards for unfortunate ships such as the *Titanic*.

Compared to land-based glaciers and ice sheets, sea ice is much more variable in distribution and thickness over annual cycles. In the polar oceans, the sea-surface freezes each winter to produce ice that extends over about 15 and 19 million km² of the Arctic and Southern oceans, respectively. Unlike glaciers and ice sheets, much of this sea ice is short-lived, with a large portion of it melting each summer to give a minimum extent of about 4 to 5 million km² in the Arctic and approximately 3 million km² in the Antarctic. The edge of the sea ice retreats poleward as the summer proceeds, with protected fjords and inlets often being the last to become clear of ice. As a result of the seasonal cycle of ice growth and melting, sea ice is usually only a few meters thick at most, as compared to hundreds or thousands of meters for glaciers and ice sheets.

A third type of ice is permafrost, which occurs in polar and high-mountain areas where the ground is permanently frozen to depths of ten to hundreds of meters. In summer, ice in the upper meter or so of the soil matrix melts to produce a soft 'active layer', which refreezes again each winter. Permanently frozen ground occupies vast areas of the Arctic beyond the margins of modern glaciers and ice sheets, including much of northern Canada, Alaska and Siberia. When taken together, almost 23 million km² of the land area of the Northern Hemisphere (approximately 5% of Earth's total surface area) is covered with permafrost. This value does not include sub-sea ancient permafrost that currently sits beneath the ocean — mostly on the extensive continental shelves north of Siberia and North America. By comparison with Arctic regions, there is relatively little permafrost in Antarctica because the ice sheet covers about 99% of that continent's land area.[1]

To understand the future evolution of Earth's cryosphere, it is instructive to look to the past. For much of Earth's four-and-a-half billion-year history, the cryosphere has been strongly influenced by climate change over various timescales. Over the past billion years, there have been six cold intervals, or ice ages, during which large ice sheets existed, intermittently, over significant parts of the planet, interspersed between extensive periods when Earth was significantly warmer than it is today. In two particularly cold periods, between about 717–660 and 650–635 million years ago, geological evidence suggests the existence of a 'Snowball Earth', when most, if not all, of the planet's surface was covered in thick layer of ice.[2] The most recent cold interval began about thirty-four million years ago when ice started to build up on the Antarctic continent, in part as a response to the opening of the deep-water Drake Passage between the Antarctic Peninsula and South America, which allowed ocean currents and winds to partially isolate Antarctica from southward heat transfer from lower latitudes. An ice sheet of varying dimensions has been present on Greenland for at least eighteen million years, while the most recent ice age in Eurasia and North America marked the beginning of the Quaternary period about 2.6 million years ago.

Earth's climate has varied during the Quaternary with a periodicity of about 100,000 years. Each 100,000 cycle can be broken down into colder (glacial) and warmer (interglacial) intervals, and the last few of these cycles are recorded in the ice itself, most notably in an Antarctic ice core over 3 km long. The sequential depth-layers of this ice core contain a frozen archive that preserves the recent climate history of Earth, including temperature and greenhouse-gas concentrations going back about 800,000 years.[3] From this record, we know that the warm period in each glacial-interglacial cycle is typically much shorter than the cold phase, making up no more than 10–20% of the cycle. Today, we are in the most recent interglacial period, which started when the Earth began warming after the last full-glacial interval about 20,000 years ago. During this last glacial maximum, the ice sheets of Antarctica and Greenland became much more extensive, and mid-latitude ice sheets built up over North America, covering Canada and reaching down as far as the Great Lakes and New York. In Europe, Scandinavia, much of western Russia and northern Britain were buried under thousands of meters of ice. At this time, global sea level was about 125 m lower than today because of the growth of these huge ice sheets on land.[4]

The past 10,000 years or so has been a time of relatively warm, interglacial climate on Earth. Ice-core records show that the early Holocene was warmer than today, and there were also colder spikes such as a brief cold event about 8,200 years ago and a cooler period known as the Little Ice Age between approximately 1400 and 1900 AD.[5] During the past century, however, and particularly over the past few decades, there has been a marked increase in Earth's air and ocean temperature. The World Meteorological Organization reported recently that global air temperatures have risen by 1.1°C since comprehensive records first became available in the mid-nineteenth century.[6] About 0.2°C of this warming has occurred in only the five years between 2010 and 2015.

The icy world, in the form of glaciers and ice sheets, sea ice and permafrost, is particularly sensitive to atmospheric- and ocean-temperature changes. Climate records show that the polar regions, and the Arctic in particular, have warmed at roughly double the global rate.[7] This so-called 'polar amplification' results from changes in surface reflective properties, in particular where melting sea ice is replaced by darker ocean water that absorbs much greater amounts of solar radiation. Computer models suggest that this 'ice-reflectance effect' (sometimes known as ice-albedo effect) will continue over the coming decades, amplifying ongoing climate warming. The exact trajectory of that warming, estimated at between less than 2° and about 5°C by 2100, will depend on the future evolution of our economic, industrial and agricultural activities, and their impact on atmospheric greenhouse gases.[8]

Over the past four decades, the availability of comprehensive satellite-based measurements has radically changed our understanding of global ice distributions. Today, we know that many glaciers around the world, and parts of the massive Greenland and Antarctic ice sheets, are already thinning and retreating as a result of atmospheric and ocean warming. Since the first Earth Day, half a century ago, glaciers in many Arctic and mountain areas have thinned by tens of meters and undergone kilometers of retreat.[9] Furthermore, the Greenland Ice Sheet has shown a clear trend towards increased melting and mass loss since the turn of the twenty-first century, with almost the entire ice-sheet surface subjected to melting in some recent summers.[10] Summer melting is now also commonplace at lower elevations in parts of the western Antarctic Peninsula, and even the 2 million km^2 West Antarctic Ice Sheet is affected, with thinning and retreat detected using very accurate satellite radar and laser altimeters.[11]

Land-based glaciers and ice sheets hold huge volumes of water, which can be released into the ocean. The loss of mass from glaciers and ice sheets thus has enormous implications for global sea level change, which affects low-lying communities world-wide.[12] This effect, together with thermal expansion of seawater resulting from recent ocean warming, is the key control on global sea level change on timescales of decades to centuries. Between 1901 and 1990, sea level rose by 1.4 mm per year. By comparison, high-accuracy satellite altimeters show that sea level rise was 3.6 mm per year from 2006 to 2015, about 2.5 times its rate over much of the twentieth century. Predictions of future sea level increases suggest a global rise of between 0.4 m and about 1 m by 2100 for low and high greenhouse-gas emission scenarios, respectively.[13] In either case, many millions of people will be displaced globally.

A second area of major concern is the decline in Arctic summer sea-ice extent, which has been monitored systematically by satellites over the past forty years.[14] September sea-ice minima have declined from around 7–8 million km² to values often less than 5 million km² over this period. For perspective, this reduction in sea-ice surface area is roughly equivalent to the land mass of India. This decline is set to continue due to ice-reflectance feedback, and computer models predict that the Arctic Ocean will be largely devoid of summer sea ice within a few decades.[15] The loss of Arctic sea ice will exert a major influence on Arctic marine ecology and the humans that depend on it, with significant geopolitical implications linked to new shipping routes and resource exploration potential.

There is another possible, and somewhat paradoxical, consequence of sea-ice decline in a warming Arctic, which is related to the effects of sea-ice formation on ocean-circulation patterns.[16] When sea ice forms, salts from seawater are rejected from the forming crystals and released into the underlying surface waters, producing very cold and salty water masses that sink to the ocean depths. In the Labrador Sea off the coast of Greenland, this deep-water formation forms one branch of a large 'ocean conveyer belt' that transports heat and nutrients southward at great depth in the North Atlantic. The upper portion of this circulation is the northward return flow of warm Gulf Stream water in the top 1,000 m or so of the North Atlantic. If the formation of deep water slows or even stops (as appears to have happened more than once in Earth's geological history), the Gulf Stream and its northward transfer of heat will also slow. This, in turn, would lead to the somewhat counter-intuitive cooling of North West Europe, or at least to a reduced warming trend.[17]

The huge areas of permafrost covering much of the Canadian and Eurasian Arctic are also vulnerable to warming. Although more difficult to measure from satellites than changing glacier extent and thickness, it appears that permafrost is responding to the enhanced Arctic warming of recent decades.[18] In most permafrost areas, ground temperatures and the rate of degradation in permafrost thickness and extent have increased over the past twenty to thirty years. This summertime melting has created challenging conditions for travel on the unstable ground, and has also begun to destabilize some built structures, such as houses and pipelines. Of potentially wider significance, a deepening of the biologically active upper layer of permafrost will increase the rate of organic matter decomposition in the soil, releasing methane to the atmosphere.[19] Methane is about thirty times more potent than carbon dioxide as a heat-trapping or greenhouse gas,[20] so methane release from the

Arctic tundra could lead to a positive-feedback loop in global warming. A similar process may be underway in the shallow ocean sediments of Arctic continental shelves, where frozen sub-sea permafrost deposits contain large quantities of frozen methane 'clathrates' that may become destabilized under warming ocean conditions.[21]

The shrinking area and volume of ice on Earth is significant not just for the polar regions, but also has important global effects, through sea level rise, ocean-circulation changes and accelerated melting of Arctic permafrost and the associated release of methane. Predictions of the rate of change in our icy world over the next few decades depend on how far humankind is prepared to curb the continued emission of greenhouse gases. The choices we make today will determine whether global temperatures increase by less than 2°C by 2100 (if we have some success in introducing and expanding alternative energy sources such as solar, wind, hydro-electric and tidal power generation), or whether the rise will be between 3° and 5°C (if only limited steps are taken to curtail greenhouse gas emissions). The icy world will respond accordingly. Having worked in the polar regions for almost four decades, I have witnessed changes in Earth's cryosphere first hand, in terms of both glacier retreat and sea-ice decline. Once glaciers are gone, there is little we can do to bring them back. Unless we take swift action to combat climate change, much of Earth's cryosphere may one day exist only as a distant memory.

Endnotes

1. D. Farinotti, M. Huss, J. J. Fürst, J. Landmann, H. Machguth, F. Maussion and A. Pandit, 'A consensus estimate for the ice thickness distribution of all glaciers on Earth', *Nature Geoscience*, 2019, 12, 168–73, https://doi.org/10.1038/s41561-019-0300-3

2. P. F. Hoffman, A.J. Kaufman, G. P. Halverson and D. P. Schrag, 'A Neoproteropzoic Snowball Earth', *Science*, 1998, 281, 1342–46, https://doi.org/10.1126/science.281.5381.1342

3. D. Lüthi, M. Le Floch, B. Bereiter, T. Blunier, J. M. Barnola, U. Siegenthaler, D. Raynaud, J. Jouzel, H. Fischer, K. Kawamura and T.F. Stocker, 'High-resolution carbon dioxide concentration record

650,000-800,000 years before present', *Nature*, 2008, 453, 379–82, https://doi.org/10.1038/nature06949

4.   K. Lambeck, H. Rouby, A. Purcell, Y. Sun and M. Sambridge, 'Sea level and global ice volumes from the Last Glacial Maximum to the Holocene', *Proceedings of the National Academy of Sciences*, 2014, 111, 15296–303, https://doi.org/10.1073/pnas.1411762111

5.   B.S. Lecavalier, D.A. Fisher, G.A. Milne, B.M. Vinther, L. Tarasov, P. Huybrechts, D. Lacelle, B. Main, J. Zheng, J. Bourgeois and A. S. Dyke, 'High Arctic Holocene temperature record from the Agassiz ice cap and Greenland ice sheet evolution', *Proceedings of the National Academy of Sciences*, 2017, 114, 5952–57, https://doi.org/10.1073/pnas.1616287114

6.   World Meteorological Organization, '2019 concludes a decade of exceptional global heat and high-impact weather', 3 December 2019, https://public.wmo.int/en/media/press-release/2019-concludes-decade-of-exceptional-global-heat-and-high-impact-weather

7.   IPPC, *Special Report on the Ocean and Cryosphere in a Changing Climate*, ed. H.-O. Pörtner et al., Geneva: IPCC, 2019, https://www.ipcc.ch/srocc/home/

8.   Ibid.

9.   A. J. Cook, L. Copland, B. P. Y. Noël, C. R. Stokes, M. J. Bentley, M. J. Sharp, R. G. Bingham and M. R. van den Broeke, 'Atmospheric forcing of rapid marine-terminating glacier retreat in the Canadian Arctic Archipelago, *Science Advances*, 2019, 5, https://doi.org/10.1126/sciadv.aau8507. See also 'Sea Level Rise, 1970–2070: A View from the Future' by Robert E. Kopp in this volume.

10.   M. McMillan et al., 'A high-resolution record of Greenland mass balance', *Geophysical Research Letters*, 2016, 43, 7002–10, https://doi.org/10.1002/2016GL069666

11.   IMBIE team, 'Mass balance of the Antarctic Ice Sheet from 2002 to 2017', *Nature*, 2017, 558, 219–22, https://doi.org//10.1038/s41586-018-0179-y.

12.   See also 'Sea Level Rise, 1970–2070: A View from the Future' by Robert E. Kopp in this volume.

13.   IPCC, *Special Report*, 2019.

14.   D. Barber et al., 'Arctic sea ice', in *Snow, Water, Ice and Permafrost in the Arctic (SWIPA),* Oslo: Arctic Monitoring and Assessment Programme, 2017, 103–36, https://www.amap.no/documents/download/2987/inline

15.   M. Wang and J.E. Overland, 'A sea ice free summer Arctic within 30 years: an update from CMIP5 models', *Geophysical Research Letters*, 2012, 39, https://doi.org/10.1029/2012GL052868.

16.   Barber et al., 'Arctic Sea Ice', 2017.

17. G. Sgubin, D. Swingedouw, S. Drijfhout, Y. Mary and A. Bennabi, 'Abrupt cooling over the North Atlantic in modern climate models', *Nature Communications*, 2017, 8, https://doi.org/10.1038/ncomms14375

18. B. K. Biskaborn et al., 'Permafrost is warming at a global scale', *Nature Communications*, 2019, 10, 264, https://doi.org/10.1038/s41467-018-08240-4

19. C. Knoblauch, C. Beer, S. Liebner, M. N. Grigoriev and E.-M. Pfeiffer, 'Methane production as a key to the greenhouse gas budget of thawing permafrost', *Nature Climate Change*, 2018, 8, 309–12, https://doi.org/10.1038/s41558-018-0095-z

20. On carbon dioxide, see 'Carbon' by David Archer in this volume.

21. M.A. Maslin, M. Owen, R. Betts, S. Day, T. D. Jones and A. Ridgwell, 'Gas hydrates: past and future geohazard?', *Philosophical Transactions of the Royal Society*, 2010, 368, 2369–93, https://doi.org/10.1098/rsta.2010.0065

Imaging Earth

—

Edward Burtynsky

Having spent close to forty years bearing witness to some of the largest extractive industries in the world, I have gained a unique perspective on the state of Planet Earth under the increasing burden of unprecedented human population growth and affluence. All living species must take from nature to survive, and we are no different. But unlike other species, there seems to be no end to our quest for food, comfort, shelter, sex — the fundamental necessities of survival that are now pursued in overdrive, far beyond our existential needs. We are compelled to progress, and have extracted resources from the land since we first stood on two feet. The entire twentieth century has been a revving up of this large consumptive engine, and this insatiable human striving has assaulted the very planet that sustains us. In a very short period of time, humankind, with its population explosion, industry and technology, has become an agent of immense global change. What this civilization leaves in the wake of its progress may be an opened and emptied Earth. But in performing these incursions, we also participate in the unwitting creation of gigantic monuments to our way of life.

My earliest understanding of deep time and our relationship to Earth's geological history came from my passion for being in nature. As a teenager, in the late 1960s, I loved to go on fishing trips, canoeing along the pristine isolated waterways of Ontario's Haliburton

https://doi.org/10.11647/OBP.0193.14

Highlands — from Kapuskasing to Cochrane and many far-flung places in between. That experience of wilderness left an enduring mark that still informs my response to landscape. I came to appreciate a state that exists without human intervention or disruption. I would paddle and look at a shoreline I knew hadn't been affected by humans, a seemingly endless expanse that marched onward with this feeling of eternity and immutability. It gave me the sense that we are just a momentary presence inhabiting this place. I think this is where I formed a perspective on our relationship to the world.

I was seven years old when I discovered my love for making art, while painting landscapes alongside my father, who painted as a hobby. I loved the tubes of oil paint and the smell of the linseed oil and the names of their colours: burnt umber, chromium blue, cadmium red. When I was eleven, I got my first camera and a complete darkroom. I immediately fell in love with photography and never looked back.

In the 1970s, I began by photographing the pristine landscape, spending a couple of years creating complex compositions of dense brush. I was honing my colour eye and my sense of how to search for compositions within chaos, learning how to find a point of view. I was learning how to see, practising scales before I could go out and compose my own songs. But I also felt that searching for the sublime in nature was an expression of nostalgia. I wanted to be true to my generation, to the world I lived in. This relevance, I decided, could be best explored by capturing images that showed how our pursuit of progress had dramatically changed the landscape.

It took a long time for my fascination with industry to manifest, though I suppose there were some early indications of this interest. I grew up in St. Catharines, Ontario, where my father worked (as I would also for a short stint) on the production line at the General Motors factory. I was able to experience, first-hand, the scale of machinery necessary to build our cars. At a young age, I saw engines being cast and steering knuckles being stamped out of red-hot ingots held with tongs by men in silver suits. The massive forge presses shook the ground and could be heard half way across the city. Once I knew what lay behind those bland brick factory walls, I would forever understand that place differently. I spent countless hours taking photographs in an abandoned factory that was full of old machinery, with enormous old windows. That early experience sent me on a life-long pursuit to record an industrial world hidden behind brick walls and barb wire fences — the industrial backbone of our economy and lifestyle.

My early professional work in the 1980s focused on capturing images of mines, quarries and rail-cuts — scars left on landscapes from the extraction and transport of Earth's raw materials to fuel our ever-growing consumption. The initial inspiration for this work was a road trip through the United States. A series of wrong turns landed me in Frackville, Pennsylvania. I remember pulling over to the side of a dirt road, and standing beside the car, surrounded by hills of coal slag. White birch trees were growing up through the black mounds, ponds were full of lime green water. It was a surreal landscape that totally destabilized me. I had never seen nature transformed on this scale before — the scale of the transformation matching only the man-made environment dominated by skyscrapers, sixty- and seventy-stories high, which I first experienced in Toronto. Frackville showed me the potential of the transformed landscape as a subject. I began to search for work that conveyed my thoughts about the industrial landscape, seeking to create images that could challenge people to consider the nature, scale and purpose of our built environments.

As my interest in mines and quarries expanded, I also began to think more about the end products of extractive resource industries. By the mid to late 1990s, the world was becoming increasingly aware of the growing accumulation of waste, unseen by many, yet massive in scale and scope. The end-products of our industrial ingenuity, scrap heaps of metal and plastic, mountains of tires and discarded computer boards, built with limited shelf lives and planned obsolescence, were becoming an increasingly important part of the human landscape. I sought to document the unseen economy working to recover raw materials from these derelict objects, wire by wire, bolt by bolt and bottle by bottle.

In 1997, the year of the ill-fated Kyoto Protocol,[1] I had what I refer to as my oil epiphany. It occurred to me that all the vast human-altered landscapes I had pursued for over twenty years had been made possible by the discovery of oil and the development of the internal combustion engine. Over the next twelve years, I researched and photographed the largest oil fields I could find. I also captured images of refineries, freeway interchanges, automobile plants and the scrap industry resulting from the recycling of cars. I began to look at motor culture, where vast tribes come together to celebrate vehicles. During this time, major changes in digital technology were rapidly opening up new ways of generating work. I could now mount and electronically control my camera from a forty-foot pneumatic monopod. By using drones, airplanes and helicopters, I could also achieve a bird's-eye perspective, rendering subjects such as transportation networks, mining, agriculture and industrial infrastructure more expansively, capturing vistas that had eluded me until then. No longer restricted by the limitations of topography or man-made structures, my lens could now literally fly.

As we entered the new millennium, my focus shifted to exploring the most fundamental aspect of life on Earth. In 2007, I began to think about water as a subject for my work, while on a production tour photographing gold mines in Australia — the first continent in this era to begin drying up. I wanted to find ways to make compelling photographs about the human systems employed to redirect and control water. Through this work, water took on a new meaning for me. I realized that water, unlike oil, is not optional. Without it we perish. Human ingenuity and the development of its industries have allowed us to control the Earth's water in ways that were unimaginable even just a century ago. While trying to accommodate the growing needs of an expanding and very thirsty civilization, we are reshaping the Earth in colossal ways.

Throughout my work, I have always been aware that lens-based media can be as subjective as other forms of art and storytelling. Yet there still remains a powerful trust of the image as evidence of our doing. I don't see myself as a reporter or as a documentary photographer, in the classic sense of the word. Rather, I see myself more as someone who is exploring ideas and trying to find images that somehow will be recognizable within a body of work produced over the arc of forty years. The American street photographer, Garry Winogrand, felt that an image succeeded when form and content were on an equal footing — one did not dominate the other. Through my selection of lens, and distance from the scene, I try to flatten the space so that the elements in the image have an equal weighting — there is no predominant object. I favour bright overcast days, so that the shadows are open, and the illumination is an even wash across the entire scene — I don't want crisp black shadows and bright highlights. I am trying to put everything into a democratic distribution of light and space across the whole field. I want everything to have an equal value so that the viewer will fall into the surface and read the detail. The compression of space through light and optics also yields an ambiguity of scale. By excluding the horizon, I can create walls of texture that challenge our ability to define spatial relationships. The work always leads me to new understandings, and new ways of perceiving.

I believe that powerful images play a small part in the huge job of raising our consciousness about who we are and the impact we are having on the world around us. The images I have captured over the past forty years are meant as metaphors to the dilemma of our modern existence; they search for a dialogue between attraction and repulsion, seduction and fear. We are drawn by desire — a chance at good living, yet we are consciously or unconsciously aware that the world is suffering for our success. But, ultimately, what I'm looking for are interesting places and moments to embody a poetic narrative of the transfigured landscape and its relationship to the industrial supply line, in an attempt to understand their meaning in our lives. We are surrounded by all kinds of consumer goods, and yet we are profoundly detached from the sources of those things. Our lifestyles are made possible by industries all around the world, but we take them for granted, as background to our existence. I feel that by showing those places, which are normally outside our experience but very much a part of our everyday lives, I can add to our understanding of who we are and what we are doing. The world is replete with subject matter.

In our new and powerful role over the planet, we have now become capable of engineering our own demise. We must come to fully appreciate the long-term consequences of our collective and individual actions. My hope is that these pictures will stimulate a process of thinking about something essential to our survival, something we often take for granted — until it's gone. For more than four decades, I have created images about the man-made transformations our civilization has imposed upon nature. As a husband and father, as an entrepreneur and provider, with a deep gratitude for his birthright in a peace loving and bountiful nation, I feel the urgency to make people aware of important things that are at stake. By describing the problem vividly, by being revelatory and not accusatory, we can help to cultivate a broader conversation about viable solutions, inspiring today's generation to carry the momentum of this discussion forward, so that succeeding generations may continue experiencing the wonder and magic of planet Earth.

Endnotes

1.   Available at https://unfccc.int/resource/docs/convkp/kpeng.pdf

Alberta Oil Sands 14

Fort McMurray, Alberta, Canada, 2007

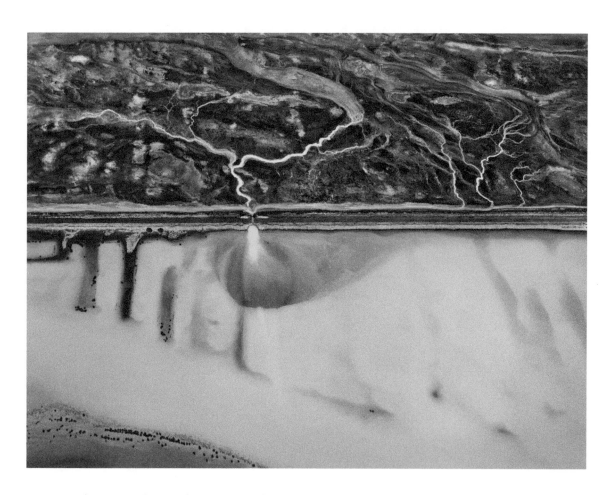

Cerro Prieto Geothermal Power Station

Baja, Mexico, 2012

Colorado River Delta 2

Near San Felipe, Baja, Mexico, 2011

Dam 6, Three Gorges Dam Project

Yangtze River, China, 2005

Photo © Edward Burtynsky, courtesy Flowers Gallery, London / Nicholas Metivier Gallery, Toronto. All rights reserved.

Dandora Landfill 3, Plastics Recycling

Nairobi, Kenya, 2016

Greenhouses

Almería Peninsula, Spain, 2010

Photo © Edward Burtynsky, courtesy Flowers Gallery, London / Nicholas Metivier Gallery, Toronto. All rights reserved.

Highway 1

Los Angeles, California, USA, 2003

Photo © Edward Burtynsky, courtesy Flowers Gallery, London / Nicholas Metivier Gallery, Toronto. All rights reserved.

Nickel Tailings 34

Sudbury, Ontario, 1996

Nickel Tailings 35

Sudbury, Ontario, 1996

Oil Bunkering 1

Niger Delta, Nigeria, 2016

Photo © Edward Burtynsky, courtesy Flowers Gallery, London / Nicholas Metivier Gallery, Toronto. All rights reserved.

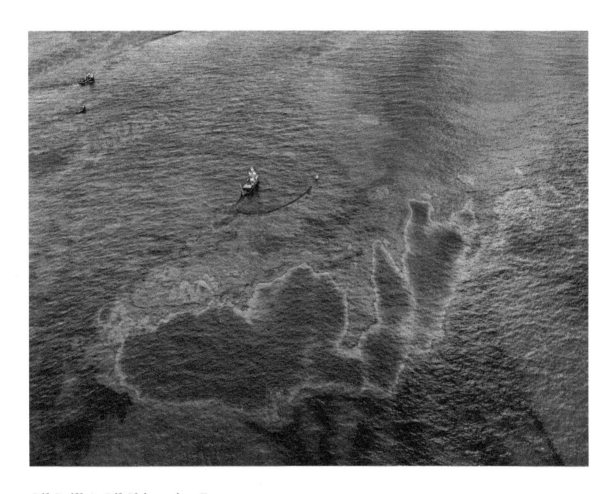

Oil Spill 4, Oil Skimming Boat

Near Ground Zero, Gulf of Mexico, June 24, 2010

Row-Irrigation, Imperial Valley

Southern California, USA, 2009

Photo © Edward Burtynsky, courtesy Flowers Gallery, London / Nicholas Metivier Gallery, Toronto. All rights reserved.

Salinas 3, Cádiz

Spain, 2013

Shipbreaking 13

Chittagong, Bangladesh, 2000

Photo © Edward Burtynsky, courtesy Flowers Gallery, London / Nicholas Metivier Gallery, Toronto. All rights reserved.

SOCAR Oil Fields 3

Baku, Azerbaijan, 2006

Uralki Potash Mine 2

Berezniki, Russia, 2017

Photo © Edward Burtynsky, courtesy Flowers Gallery, London / Nicholas Metivier Gallery, Toronto. All rights reserved.

Mother Earth

—

Deborah McGregor

Long before the first Earth Day in 1970, Indigenous peoples around the globe developed complex knowledge systems that facilitated sustainable relationships with the natural world. These Indigenous knowledge systems (IKS) have been utilized, transformed and innovated by Indigenous peoples to sustain their communities, territories and Nations since time immemorial, and passed down over countless generations. Though highly diverse in nature, IKS around the world share certain common philosophical foundations, including a responsibility to maintain and enhance relationships with 'Mother Earth' as a living entity, and a profound connection with Earth's natural systems that is acknowledged every day. In the words of the Indigenous scholar and activist, Daniel Wildcat, every day is Earth Day from an Indigenous perspective.

Indigenous knowledge systems exist in various forms under different names, including 'local knowledge', 'folk knowledge', 'people's knowledge', 'traditional wisdom', 'ethnoscience', 'native science', 'traditional science', 'traditional knowledge' (TK) and 'traditional ecological knowledge' (TEK). Yet, all these ways of understanding the world are united in their challenge of the dominant political and economic world order, and their calls for fundamental change to achieve sustainability for all beings on Mother Earth. Indigenous activist Winona LaDuke describes IKS as 'the culturally and spiritually based way in which Indigenous peoples relate to their ecosystems', through a 'way of living or

    https://doi.org/10.11647/OBP.0193.15

being' in the natural world.[1] IKS is about how one *relates to* Mother Earth, rather than the information *gathered from* Mother Earth through other forms of knowledge.

In the face of rapid climate change and a myriad of other human pressures on Mother Earth, it is increasingly clear that western knowledge systems have thus far failed to achieve sustainability; science and technology alone cannot get us out of our current crises. Other approaches are needed, and it is logical that IKS comes to be seen as a relevant and viable system for understanding our present situation, and providing a basis from which to work toward solutions. IKS has applications in many fields, including law, governance, social work, health and medicine, philosophy, education and environment. Anyone who is truly interested in sustainable relationships with the Earth should also be interested in IKS. But this realization has come late to colonial societies.

On Earth Day, fifty years ago, Indigenous voices and perspectives were conspicuously absent. Yet, Indigenous peoples have long been calling for the recognition of IKS in the decision-making processes that impact their lives, lands and waters. It is only within the past few decades that the United Nations has recognized Indigenous voices. This recognition of IKS on the global stage coincided with the increasing assertion by Indigenous people of their rights, the recognition of these rights by the international community, and the growing understanding of Indigenous rights, IKS and environmental sustainability as interwoven concepts.

International recognition of the value of IKS in addressing global environmental issues goes back to the early 1980s, when the International Union for the Conservation of Nature (IUCN) established a Working Group on Traditional Ecological Knowledge, or TEK.[2] This early international intervention was supported by a series of workshops and symposia examining the value of TEK for natural resource management, and the unique perspectives of Indigenous knowledge systems on various environmental issues. The 1987 Brundtland Report of the World Commission on Environment and Development emphasized the important role of Indigenous peoples in sustainable development, and served as a catalyst for increased recognition of IKS worldwide.[3] This landmark document not only introduced the concept of *sustainable development* to mainstream discourse, but also provided international recognition of the potentially vital contribution of Indigenous peoples to global environmental stewardship. This represented a significant shift in the public dialogue on Indigenous environmental issues — from marginalized and vulnerable peoples in need of (sustainable) development, to cultures with millennia of experience living sustainably on the land.

Five years later, at the 1992 United Nations Conference on Environment and Development (the Earth Summit), held in Rio de Janeiro, the legally binding Convention on Biodiversity (CBD) was signed.[4] The CBD and Agenda 21[5] emerged as two of the most significant outcomes of the Earth Summit, setting out international commitments for maintaining the planet's ecosystems. These landmark agreements, signed by a majority of the world's governments, reiterated the important role of Indigenous people and their knowledge for achieving environmentally sustainable development. Both Agenda 21 and the CBD formally acknowledged the historical relationships of Indigenous peoples to their lands, and the wealth of traditional ecological knowledge developed over many generations. At this time, Indigenous peoples generated their own agenda on the international stage at the World Conference of Indigenous Peoples on Territory, Environment and Development, which was held in conjunction with the Earth Summit. A key outcome of this meeting was the Indigenous Peoples Earth Charter (part of the Kari-Oca Declaration)[6] which stated that, 'Recognizing indigenous peoples' harmonious relationship with Nature, indigenous sustainable development strategies and cultural values must be respected as distinct and vital sources of knowledge'. More directly, as stated in section 98 of the Indigenous Peoples Earth Charter (Kari-Oca Declaration), 'Traditional knowledge has enabled indigenous peoples to survive'.

The potential role of Indigenous knowledge in achieving global sustainability was reaffirmed a decade later at the 2002 World Summit on Sustainable Development (WSSD), held in Johannesburg, South Africa. As with the 1992 Rio Summit, Indigenous peoples held their own parallel summit, which generated the Kimberley Declaration of the International Indigenous Peoples Summit on Sustainable Development.[7] The Kimberley Declaration states that, 'Today we reaffirm our relationship to Mother Earth and our responsibility to coming generations to uphold peace, equity and justice', and that, 'Our lands and territories are at the core of our existence — we are the land and the land is us'. This worldview was further articulated at Rio+20 (The United Nations Conference on Sustainable Development), held in Rio de Janeiro in 2012, twenty years after the original Earth Summit. One result of this 2012 meeting was the Kari-Oca 2 Declaration, which included a call for the international community to 'recognize the traditional systems of resource management of the Indigenous Peoples that have existed for the millennia, sustaining us even in the face of colonialism' (Kari-Oca 2 Declaration 2012).[8]

The Kari-Oca 2 Declaration, and others, highlight the growing recognition of Indigenous voice and perspective on the global stage. Indeed, since the first 'Earth Summit' in 1992, the United Nations has increasingly promoted global recognition of traditional knowledge systems in achieving various environmental goals. This support has taken the form of intergovernmental guidance for the use, protection, access and sharing of traditional knowledge, its potential as a complement to science, and the need for on-the-ground support to ensure its continued innovation and vitality. In many respects, these trends over the past four decades represent opportunities for the involvement of Indigenous peoples in addressing environmental challenges locally, regionally and globally.

More recent agreements, including the United Nations Millennium Development Goals (2000), have promoted a human and Indigenous rights approach to development, including the recognition of IKS. This represents an explicit acknowledgement of the unique role that Indigenous peoples can play in achieving sustainable development. Indeed, the recent 2030 Agenda for Sustainable Development, adopted by the United Nations in 2015, offers opportunities for Indigenous peoples and their knowledge systems to participate directly in global environmental governance. Thus far, however, this opportunity has yet to be fully realized.

Internationally, one of the most important undertakings in recent years has been the adoption, in 2007, of the United Nations Declaration on the Rights of Indigenous Peoples (UNDRIP),[9] following decades of advocacy by Indigenous peoples. UNDRIP explicitly recognizes the importance of Indigenous knowledge as having a key role in realizing a sustainable self-determined future. Indigenous philosophies are also becoming increasingly evident in various international Indigenous declarations pertaining to the environment, most notably the Universal Declaration on the Rights of Mother Earth developed at the World People's Conference on Climate Change in Cochabamba, Bolivia, in 2010.[10] Article 1 of the Declaration states that 'Mother Earth is a living being'. Article 3 states that 'Every human being is responsible for respecting and living in harmony with Mother Earth'. This offers a long-range perspective of how Mother Earth is understood, and the necessity of protecting her rights in order to sustain humanity. At its core, the declaration recognizes that humans are part of Mother Earth, 'an indivisible, living community of interrelated and interdependent beings with a common destiny'.

Despite growing awareness and recognition of the value of IKS in addressing local, regional and global environmental challenges, there is still much more that can and must be done. Indigenous interventions, as expressed through numerous declarations over past decades, offer a path towards an alternative future, based in part on the concept of *Buen Vivir*, Living Well, with the Earth. Indigenous peoples who gathered at the 2012 Rio+20 conference offered an alternative pathway to the unsustainable approaches proposed by international and state actors. The declaration that emerged from this meeting challenged the international community to embrace a new approach to sustainable developed, informed by a deep-rooted respect for Earth's natural systems:

> Indigenous peoples call upon the world to return to dialogue and harmony with Mother Earth, and to adopt a new paradigm of civilization based on *Buen Vivir* — Living Well. In the spirit of humanity and our collective survival, dignity and well-being, we respectfully offer our cultural world views as an important foundation to collectively renew our relationships with each other and Mother Earth and to ensure that *Buen Vivir* / living well proceeds with integrity.[11]

Buen Vivir calls for an expanded view of community, where balanced relationships are sought between humans and other entities in the natural world (animals, plants, birds, forests, waters, etc.), as well as with future generations. To live well with the Earth, humanity must recognize the agency of Mother Earth.

Buen Vivir is more than a philosophy. It is way of life, a responsibility to live in a way that supports the well-being of Mother Earth as expressed in the 2013 Lima Declaration, which emerged from the World Conference of Indigenous Women: Progress and Challenges Regarding the Future We Want:

> Protection of Mother Earth is a historic, *sacred and continuing responsibility of* the world's Indigenous Peoples, as the ancestral guardians of the Earth's lands, waters, oceans, ice, mountains and forests. These have sustained our distinct cultures, spirituality, traditional economies, social structures, institutions, and political relations from immemorial times. Indigenous women play a primary role in safeguarding and sustaining Mother Earth and her cycles.[12]

Justice for the Earth, as expressed in this way by Indigenous peoples, conveys a distinct path forward and a vision that includes all life.

Whatever one's viewpoint, Indigenous peoples' continued assertions of the rights of Mother Earth can no longer be seen as simply philosophical reflections or ancient history. On the contrary, they are becoming a reality in certain state legal systems. Emerging conceptual frameworks such as Earth jurisprudence, Earth justice and wild law are gaining currency and increasingly becoming the topic of much debate. Such Earth-centred legal philosophies emphasize the interconnections and interdependence of humanity and the natural world. The conceptual frameworks that uphold the rights of Mother Earth are gaining ground, and have been enacted through constitutional and legal mechanisms in both Ecuador and Bolivia. In 2008, Ecuador adopted specific mention of the rights of Mother Earth into its Constitution. Bolivia adopted the Law of the Rights of Mother Earth (2010), which outlines principles that seek harmony with Mother Earth, along with obligations and duties of the state and its people to protect and uphold these rights. In another recent example from New Zealand, the Whanganui River *iwi* (tribe) entered into an agreement with the Crown to recognize the Whanganui River as a living and legal entity.

These innovative pathways to environmental sustainability are based in part on ancient philosophies. They reflect the persistence of Indigenous peoples' influence, and their role in creating an expanded dialogue of sustainability informed by their understanding of Mother Earth and humanity's obligations to her. The ideas are both ancient and innovative.

If we let them guide us into the future, perhaps humanity will one day celebrate Earth Day with Mother Earth herself.

Endnotes

1. W. LaDuke, 'Traditional ecological knowledge and environmental futures', in *Endangered Peoples: Indigenous Rights and the Environment*, ed. Colorado Journal of International Environmental Law and Policy, Niwot, CO: University Press of Colorado, 1994, 127–48, at 128, https://www.uky.edu/~rsand1/china2017/library/LaDuke.pdf

2. N. Williams and G. Baines, *Traditional Ecological Knowledge: Wisdom for Sustainable Development*, Canberra: Centre for Resource and Environmental Studies, Australian National University, Australia, 1993, https://journals.uair.arizona.edu/index.php/JPE/article/view/20159, https://doi.org/10.2458/v2i1.20159

3. World Commission on Environment and Development, *Our Common Future*, Oxford: Oxford University Press, 1987, http://www.un-documents.net/our-common-future.pdf

4. Available at http://www.cbd.int/convention/text/default.shtml

5. Available at http://sustainabledevelopment.un.org/content/documents/Agenda21.pdf

6. Available at https://www.dialoguebetweennations.com/IR/english/KariOcaKimberley/intro.html

7. Available at https://www.dialoguebetweennations.com/IR/english/KariOcaKimberley/Kimberley Declaration.htm

8. Available at https://www.ienearth.org/kari-oca-2-declaration/

9. Available at https://www.un.org/development/desa/indigenouspeoples/wp-content/uploads/sites/19/2018/11/UNDRIP_E_web.pdf

10. Available at https://therightsofnature.org/universal-declaration/

11. Available at http://www.forestpeoples.org/sites/default/files/publication/2012/06/final-political-declaration-adopted-rio20-international-conference-indigenous-peoples-self-determina.pdf

12. Available at https://www.unido.org/sites/default/files/files/2018-12/UNIDO_GC15_Lima_Declaration.pdf

Sea Level Rise, 1970–2070
A View from the Future

—

Robert E. Kopp

Author's Note

The initial draft of this manuscript mysteriously appeared in my cloud drive on Earth Day 2020, with a time stamp fifty years later, April 22, 2070. It appears to have been written for a volume similar to this one, commemorating the Earth Day centennial. The description of the state of scientific knowledge through 2020 is accurate, and the discussion of future sea level and ice-sheet changes, as well as the challenges of adapting to rising sea level, are likewise consistent with current understanding. I have added endnotes, where appropriate, to support these descriptions. I cannot vouch for the accuracy of the depiction of the specific events of the next half-century, but leave these intact; they seem to me to represent one plausible future.

One hundred years ago, at the time of the first Earth Day, in 1970, sea level rise was not a pressing global concern. While scientists had already been studying sea level change for over a century, it was mostly an intellectual curiosity. Nineteenth-century geophysicists had calculated how the growth and shrinkage of ice sheets reshape Earth's

https://doi.org/10.11647/OBP.0193.16

gravitational field, affecting sea level differently in different places. Early twentieth-century oceanographers had identified numerous processes leading to short-term variations in local sea level. In 1941, stitching together data from tide gauges around the world, Caltech geophysicist Beno Gutenberg identified a global average sea level rise of about 1.1 mm per year over the preceding half-century. But it wasn't until about three decades later that widespread scientific and public concern about a potential rapid acceleration of global sea level rise began.

Writing in the journal *Nature* in 1978, John Mercer, a glaciologist at Ohio State University, sounded an alarm about the potential melting of the West Antarctic Ice Sheet (WAIS) in response to modest levels of warming. The WAIS, he noted, sits with its base largely below sea level, and is buttressed by floating ice shelves around its perimeter; these factors make it potentially unstable in the face of the warming ocean and air. While the scientific understanding of ice-sheet physics has evolved, Mercer correctly identified the broad scope of the hazard we now see playing out, nearly a century later. He described the deglaciation of West Antarctica as a potentially 'disastrous result of continued fossil fuel combustion' that could lead to 'major dislocations in coastal cities, and submergence of low-lying areas such as much of Florida and the Netherlands'.[1] Indeed, at the turn of the twenty-first century, the WAIS contained enough water to raise global average sea level by 4.5 m, and there was active debate over whether some of the 3.5 m sea level equivalent of ice that sits below the ocean surface was already committed to collapse over the next couple centuries. Today, collapse is clearly underway in multiple sectors of the ice sheet.[2]

Fortunately, the planet's two other major ice sheets are so far proving more stable. The East Antarctic holds 53 m sea level equivalent of ice; of this, 19 m sea level equivalent sit with their base below the ocean surface and are potentially vulnerable to the same instabilities playing out in West Antarctic. But so far this ice sheet has only shrunk by a few centimeters and does not seem in imminent danger of collapse. The Greenland Ice Sheet contains enough water to raise sea level by 7.4 m. While geological records from past warm periods suggest we may already have warmed the planet enough to lose a substantial chunk of this ice sheet, it currently appears that its loss will take many millennia.[3]

While the ice sheets are the major (and most visible) driver of sea level change today, they aren't the only important factor. Indeed, as of the fiftieth Earth Day in 2020, they weren't even the dominant one. From 1993 — when the first satellite providing global sea level observations was launched — to 2020, global average sea level rose by about 8 cm. Of that 8 cm, about 40% was due to the thermal expansion of ocean water as it warmed, and another quarter to melting mountain glaciers.[4] The remaining 35% was due to accelerating ice losses from both Greenland and West Antarctica.[5]

That was the global story. Around the world, however, for a variety of reasons, different places experienced different rates of sea level change. For one thing, surface winds and ocean currents are important drivers of local sea level changes — indeed, the dominant driver on a year-to-year basis. Other factors also come into play. Over the twentieth century, many inhabited river deltas, such as the Mississippi Delta in Louisiana, experienced sea level rise several times greater than the global average. In these areas, which rest upon loosely consolidated sediments, the weight of the sediments can lead to a sinking of the land surface, and thus a relative sea level rise — a process accelerated when humans pump water, oil or gas out from between the sands. Conversely, other areas — such as parts of Alaska near melting glaciers — actually experienced a relative *drop* in sea level over the twentieth century, as a result of various geophysical processes. When a glacier or ice sheet melts, it exerts a weaker gravitational pull on nearby water bodies, leading to a local drop in sea level, and enhanced sea level rise farther away. The loss of glaciers also leads to a gradual 'rebound' of Earth's crust and mantle underneath the reduced load, further contributing to the relative drop in sea level. Thus, the actual changes in sea level experienced at any place on Earth can differ quite significantly from overall global trends.

Even back in 2020, sea level rise was having real — and increasingly costly —impacts. High-tide floods had become increasingly common in many areas. In coastal New Jersey, for example, impactful high-tide flooding increased from about one day in a typical year in the 1980s to more than five days in a typical year in the 2010s, with increases of similar magnitude occurring along much of the US Atlantic Coast.[6] Such increasingly frequent floods were starting to impact commerce, farming, property markets, sanitation and groundwater supplies in low-lying coastal areas around the world. In addition, coastal storms were causing more impactful flooding. During 2012's Hurricane Sandy, for example, higher seas meant that floodwaters reached the homes of more than 80,000 people who would not otherwise have been affected. Yet, adaptation efforts were quite limited: for example, in New Jersey in the five years after Hurricane Sandy, state-funded buyouts of flood-prone properties were outnumbered about five-fold by new houses built in the future flood plain.

From the vantage point of Earth Day 2020, the next couple decades of sea level rise were already set in motion, with the world almost unavoidably set for a global average rise of about 7 to 17 cm, sufficient to make high-tide flooding a multiple-weeks-per-year affair in many coastal communities.[7] Beyond the 2040s, however, there were two major drivers of sea level uncertainty: the future trajectory of global greenhouse gas emissions and ice-sheet physics. The former was a topic of urgent policy and diplomacy; the latter, a scientific story slowly being unveiled.

Had the world stuck to the aggressive temperature targets laid out in the Paris Agreement (drafted in 2015 and signed in 2016),[8] the additional global sea level rise over the last half-century (from 2020 to 2070) would most likely have been about 30 cm. The odds were quite good — about nine chances in ten, based on computer models — that sea level rise would have stayed below 50 cm.[9]

Unfortunately, the Paris Agreement was a very limited success. To be sure, intensified efforts in the 2020s did stabilize global greenhouse emissions, getting the world off the 'business-as-usual' path of growing fossil fuel consumption. But it took until the 2050s for global emissions to really start falling. In addition, an unexpected and rapid reduction in tropical cloud cover has accelerated the global temperature rise, with global average temperatures in 2070 now closing in on 3°C above nineteenth-century levels. The Paris Agreement's 1.5 and 2.0°C temperature targets are now a distant memory.

The world's lethargic approach to mitigation wasn't enough to slow cascading instabilities in the West Antarctic ice sheet. Global average sea level rise since the year 2000 has exceeded 60 cm, and it seems on track to double that by the end of the twenty-first century. By the 150th Earth Day in 2120, we may be nearing 2 m of sea level increase.[10]

And we may see even more dramatic changes over the longer term. Computer models and geological records of past warm periods suggest that 3°C of global warming would lead to about 10 m of total rise over the next two millennia if the planet were left to its own devices.[11] Such a drastic increase in global sea level would flood 2.6 million km² of habitable land surface — an area currently home to over 10% of the global population.

But it looks increasingly unlikely that the planet will be left to its own devices. While mid-century proposals from India and the Alliance of Small Island States to engineer the planet's climate with stratospheric aerosol pollution seem to have been quieted by threats from China's risk-averse leadership, efforts to artificially remove carbon dioxide from the atmosphere have seen rapid growth in the last decade.[12] It's quite possible that, by Earth Day 2120, the rate of deliberate removal of atmospheric carbon dioxide will match the mid-century rate of human emissions. If such efforts can be sustained, Earth's temperature may cool back to near its pre-industrial levels by the first half of the twenty-third century, with global sea level rise slowing to a more measured pace by the twenty-fourth century. Computer models suggest that such sustained emissions reductions efforts might be sufficient to keep the long-term global average sea level rise below 4 m.[13]

While the delayed benefits of climate change mitigation have yet to be seen by the world's coastal communities, there have been greater successes in adapting to the effects of sea level rise. In the once-laggard United States, for example, the 2020s saw some major steps forward. Following the devastating New Orleans flooding of 2023 — when levees built inadequately in the aftermath of Hurricane Katrina were overtopped during Hurricane Louis — the US president's major adaptation initiatives became one of the few elements of her proposed Green New Deal to win broad bipartisan support.

The centerpiece of these adaptation efforts was the Community And Regional Climate Adaptation Act (CARCAA) of 2024. This legislation provided federal support for states and public universities to work with communities in developing formal adaptation strategies, known as Community Adaptation Pathways (CAPs), and coordinating these strategies within and across state lines. CARCAA also set up an Adaptation Trust Fund to help finance the implementation of the CAPs, providing multi-year budgetary stability that was often missing in infrastructure projects in the US, and allowing lower-income communities to prepare in the same deliberate fashion as wealthier areas.

In coastal communities, each CAP considers four basic adaptations to sea level rise — accommodating more frequent flooding, defending against incoming waters, advancing into the ocean by elevating and extending the land, and relocating to safer ground. The CAPs evaluate the levels of sea level rise at which these four options would work, their financial costs, the resulting change in a community's risk profile, and the extent to which implementation would enable or hinder other options. And, importantly, the CAP must consider when and where relocation is the most viable remaining option.

Accommodation to more frequent tidal- and storm-driven flooding serves as the first rung of a coastal CAP, and it has taken many forms. At its most basic, accommodation requires improving disaster response: for example, enhancing emergency communications and ensuring transportation networks function well during evacuations. Historical experience shows that mutual aid among the affected people plays a critical role in the immediate aftermath of a disaster, and large-scale preparedness exercises, modeled on the Great California ShakeOut, are one of the lasting legacies of CARCAA. Accommodation also includes physical changes to buildings and infrastructure, including traditional approaches like raising infrastructure, elevating buildings and wet-proofing basements to tolerate occasional flooding. More innovative approaches have also been developed, such as creating buildings that can safely float during a flood. In the aftermath of 2025's Hurricane Tanya, the rebuilding of the Naval Academy and the Naval Station Norfolk led to major innovations in this area, the legacy of which can be seen today across the world, from Washington, DC, to Guangzhou.

Coastal defense often takes the form of hard infrastructure: surge barriers that can be closed in the event of incoming high waters, as well as permanent levees and flood walls. It can also include softer infrastructure, such as periodically replenished beach dunes. Oyster reefs and salt marshes also provide substantial protection against waves, although they are generally less effective in protecting against longer-lasting storm surges and tides. Defensive structures that combine hard and soft elements in cities like New York and Boston have become a signature element of Green New Deal-era architecture.

Coastal advance involves reclaiming land from the sea and building it up to higher elevation. There are many historic examples. In the Netherlands, half of the country's land area is composed of polders, low-lying areas reclaimed over the course of centuries from marshland. In the aftermath of the Great Seattle Fire of 1889, streets that were originally near the tidal zone were elevated by about 12 ft, with the original first floors of buildings turned into basements. More recently, in the late twentieth and early twenty-first centuries, Shanghai increased its land area by over 6% by moving sand and using sea walls to capture sediment carried by the Yangtze River.

No defensive structures are failsafe, and in a world of rising sea levels, neither defense nor advance can be safely used without accompanying measures to accommodate flooding. Indeed, coastal defense and advance can create a false sense of security, allowing populations to continue to grow in areas with substantial flood exposure, as was demonstrated by the flooding of Lower Manhattan during Hurricane Susan in 2043. Twenty years earlier, following a pre-CARCAA plan developed by Mayor Bill de Blasio, New York City extended the southern portion of Manhattan to make space for new flood protection structures. But the condo developers followed quickly thereafter, putting more people in a vulnerable area. In the aftermath of Hurricane Rebekah, skepticism about coastal advance has significantly restricted its role in modern CAPs.

The fourth adaptation option — relocation — moves people out of harm's way. Unplanned relocation is often associated with the aftermath of a disaster. For example, New Orleans permanently lost about 15% of its population in the aftermath of Hurricane Katrina in 2005, and another 10% in the aftermath of Hurricane Lee in 2023. But perhaps the biggest success of CARCAA and follow-on measures was the widespread community deliberations about possible future relocation in the development of the CAPs. With gradual, community-driven transitions now mapped out in advance for vulnerable areas, there has been no large-scale disaster-driven population displacements in the US in over four decades.

A key to the success of the CAP process is that it is not imposed by the federal or state government; rather, these higher levels of government participate in a supporting role, providing funding for the process and incentives for participation, as well as identifying (and occasionally removing) barriers that might prevent options from playing out. Often, public universities — generally much more highly trusted than federal or state government, especially early on in the CARCAA era — play critical roles both as conveners and as sources of expert knowledge. The CAPs demonstrated that community voices could have real impact. They showed that communities, universities and higher levels of government working together could limit some of the damage of climate change's increasingly severe effects, and even create beauty in new public works.

Even so, efforts to cope with intensifying coastal flooding around the world have been less successful in lower-income countries. When freak hurricanes struck West Africa in the 2050s, millions of people were dislocated, and a substantial fraction of them sought refuge in the European Union. A similar wave of migration into the EU was kindled by Hurricane Milton's fierce landfall in England in 2062. With last year's Cyclone Kyarr sparking a similar dislocation in Myanmar and forcing the issue of disaster migration onto the Chinese agenda, it seems increasingly likely that a long-overdue examination of international migration law will be a key part of the global adaptation agenda in the 2070s.

Endnotes

1. J. H. Mercer, 'West Antarctic ice sheet and CO_2 greenhouse effect: a threat of disaster', *Nature*, 1978, 271, 321–25 at 325, https://doi.org/10.1038/271321a0

2. See 'Ice' by Julian Dowdeswell in this volume.

3. P. Fretwell et al., 'Bedmap2: improved ice bed, surface and thickness datasets for Antarctica', *The Cryosphere*, 2013, 7, 375–93, https://doi.org/10.5194/tc-7-375-2013; M. Morlighem et al., 'BedMachine v3: Complete bed topography and ocean bathymetry mapping of Greenland from multibeam echo sounding combined with mass conservation', *Geophysical Research Letters*, 2017, 44, 11,051–11,061, https://doi.org/10.1002/2017GL074954

4. See 'Ice' by Julian Dowdeswell in this volume.

5. WCRP Global Sea Level Budget Group, 'Global sea-level budget 1993–present', *Earth System Science Data*, 2018, 10, 1551–90, https://doi.org/10.5194/essd-10-1551-2018

6. W. Sweet, G. Dusek, J. Obeysekera and J. J. Marra, *Patterns and Projections of High Tide Flooding Along the US Coastline Using a Common Impact Threshold*, National Oceanic and Atmospheric Administration, Silver Spring, Maryland: US Department of Commerce, 2018, 56, https://tidesandcurrents.noaa.gov/publications/techrpt86_PaP_of_HTFlooding.pdf

7. B. P. Horton, R. E. Kopp, A. J. Garner, C. C. Hay, N. S. Khan, K. Roy and T. A. Shaw, 'Mapping sea level change in time, space and probability', *Annual Reviews of Environment and Resources*, 2018, 43, 481–521, https://doi.org/10.1146/annurev-environ-102017-025826

8. Available at https://unfccc.int/resource/docs/2015/cop21/eng/l09r01.pdf

9. J. L. Bamber, M. Oppenheimer, R. E. Kopp, W. P. Aspinall and R. M. Cooke, 'Ice sheet contributions to future sea level rise from structured expert judgment', *PNAS*, 2019, 116, 11195–200, https://doi.org/10.1073/pnas.1817205116

10. P. U. Clark et al., 'Consequences of twenty-first-century policy for multi-millennial climate and sea level change', *Nature Climate Change*, 2016, 6, 360–69, https://doi.org/10.1038/nclimate2923

11. R. M. DeConto and D. Pollard, 'Contribution of Antarctica to past and future sea level rise', *Nature*, 2016, 531, 591–97, https://doi.org/10.1038/nature17145

12. See 'Air' by Jon Abbatt in this volume.

13. D. Ehlert and K. Zickfeld, 'Irreversible ocean thermal expansion under carbon dioxide removal', *Earth System Dynamics,* 2018, 9, 197–210, https://doi.org/10.5194/esd-9-197-2018; P. J. Applegate and K. Keller, 'How effective is albedo modification (solar radiation management geoengineering) in preventing sea level rise from the Greenland Ice Sheet?', *Environmental Research Letters,* 2015, 10, 084018, https://doi.org/10.1088/1748-9326/10/8/084018

Climate Negotiations

—

Rosemary Lyster

In the fifty years since the first Earth Day on 22 April, 1970, the planet has been irrevocably changed. No matter the number of international studies and reports that have predicted the crisis since that day, nothing has arrested the steady decline of ecosystems and natural resources. Humans and non-humans alike now face threats that take them beyond their coping range and resilience. The most recent Special Reports of the Intergovernmental Panel on Climate Change (IPCC) provide clear and consistent warnings that climate change is happening and happening fast. Global warming is likely to reach 1.5°C, or even 2°C, between 2030 and 2050 if greenhouse gas emissions are not reduced, significantly increasing the risk of 'long-lasting or irreversible changes.'[1] Meanwhile, climate change has already adversely impacted vulnerable terrestrial ecosystems, while also contributing to desertification, land degradation[2] and significant changes to the oceans and the cryosphere.[3] At the same time, the 2019 Global Assessment Report on Biodiversity and Ecosystem Services, a report by the UN's Intergovernmental Science-Policy Platform on Biodiversity and Ecosystem Services (IPBES), now warns that biodiversity is declining faster than at any time in human history with around one million species already facing extinction, many within decades, unless action is taken to reduce the intensity of drivers of biodiversity loss.[4]

So, what has the international community been doing about all of this in the previous five decades? Perhaps not coincidentally, the first coordinated response to the looming

     https://doi.org/10.11647/OBP.0193.17

environmental crisis occurred shortly after the first Earth Day, when the United Nations convened the 1972 Stockholm Conference on the Human Environment. This was the first time that world leaders had gathered specifically to address global environmental issues, and their work led to the adoption of the Declaration of the United Nations Conference on the Human Environment (the Stockholm Declaration).[5] It was here that nations acknowledged that 'man has acquired the power to transform his environment in countless ways and on an unprecedented scale'. Critically, Principle 1 of the Declaration also heralded the advent of a right to environment. It stated:

> Man has the fundamental right to freedom, equality and adequate conditions of life, in an environment of a quality that permits a life of dignity and well-being, and he bears a solemn responsibility to protect and improve the environment for present and future generations.

So began a new dawn of environmental awakening and a commitment by all nations to acknowledge the essential role that Earth plays in sustaining human existence. Principle 1 has spawned the inclusion of environmental rights in over one hundred Constitutions around the world. It seemed that international environmentalism would enjoy a 'golden age' as governments collaborated through multilateral institutions to protect the planet.

But subsequent progress was slow. Despite the bold assertions of the Stockholm Declaration, it would take many years for the global community to acknowledge the threat that climate change, in particular, posed to the planet. A first important step was the 1987 publication of Report of the World Commission on Environment and Development: *Our Common Future*.[6] This document, also known as the Brundtland Report, was the outcome of an independent political and scientific commission led by former Norwegian Prime Minister Gro Harlem Brundtland, and commissioned by the World Commission on Environment and Development. It responded to an urgent call by the General Assembly of the United Nations 'to propose long-term environmental strategies for achieving sustainable development by the year 2000 and beyond'. The Commission provided a definition of sustainable development which changed the face of environmental regulation forever:

> Sustainable development is development that meets the needs of the present without compromising the ability of future generations to meet their own needs.

A rereading of the Brundtland Report, more than three decades after its publication, is a rather poignant reminder of the pioneering work, and even optimism, of this Commission as it grappled with the major social, economic, environmental and geo-political issues of the day, while envisioning a different future. Yet the authors also presented a clear warning that 'the time has come to take the decisions needed to secure the resources to sustain this and coming generations'.

Around the time that the Brundtland Commissioners were undertaking their work, another important development occurred. Faced with growing evidence of global climate change, the World Meteorological Organization (WMO) established the Intergovernmental Panel on Climate Change (IPCC) as a definitive international scientific body to advise the United Nations on the state of Earth's evolving climate. Anyone who has read the IPCC's periodic Assessment Reports[7] will know that each successive Report expresses increasing degrees of confidence about the observed changes in the global climate, as well as the model-based predictions for the future.

The first IPCC report appeared in 1990, just a few years after the publication of the Brundtland Report. Together, these two documents were a significant driving force for the 1992 United Nations Conference on Environment and Development (the Rio Conference). It was here that the United Nations Framework Convention on Climate Change (UNFCCC) was born, as part of a package of measures for the twenty-first century, including the Rio Declaration on Environment and Development,[8] Agenda 21,[9] the Convention on Biological Diversity (CBD),[10] and the Forest Principles.[11] The Preamble to the UNFCCC[12] contains the following principles, which resonate with the underlying norms of International Law as well as sustainable development: that the Earth's climate and adverse effects are a common concern of humankind; that the greenhouse effect will warm Earth's surface and atmosphere and adversely affect natural ecosystems and humankind; that there is a need for an appropriate international response in accordance with common but differentiated responsibilities; that developed countries have a historical but also current responsibility for their emissions, while emissions originating in developing countries will need to grow in future; that developed countries should take immediate action to develop comprehensive strategies; and that responses to climate change should be coordinated with social and economic development. Pertinently, the Parties acknowledged that low-lying small island developing states, and other developing countries prone to floods, drought and desertification, are particularly vulnerable to the adverse effects of climate change.

The principal objectives of the UNFCCC and Rio Declaration serve as a reminder that in 1992 there was international agreement to establish 'a new and equitable global partnership' and to develop international agreements which would 'respect the interests of all and protect the integrity of the global environmental and developmental system'. Perhaps the most influential elements of the Rio Declaration have proved to be the principle of intergenerational equity, the precautionary principle and the polluter pays principle. Intergenerational equity requires current rates of development to equitably meet the development and environmental needs of present and future generations. The precautionary principle holds that, 'where there are threats of serious or irreversible damage, lack of full scientific certainty shall not be used as a reason for postponing cost-effective measures to prevent environmental degradation'. Finally, the polluter pays principle envisages the 'internalisation of environmental costs and the use of economic instruments, taking into account the approach that the polluter should, in principle, bear the cost of pollution'.[13]

The ultimate objective of the UNFCCC was to achieve stabilization of greenhouse gas (GHG) concentrations in the atmosphere at a level that would prevent dangerous anthropogenic interference with the climate system. Such a level was to be achieved within a timeframe (not clearly articulated in 1992) sufficient to allow ecosystems to adapt naturally to climate change and to ensure sustainable food production and economic development. The first step towards legally-binding GHG emissions targets was the 1997 Kyoto Protocol, whereby developed countries agreed that overall emissions would be capped at 5% below 1990 levels by the end of 2012.[14] Developing countries were not required to meet any targets, and this was seen by some nations as a major point of contention.

With the early focus on reducing GHG emissions in the mid-1990s, there was a view that identifying climate change adaptation options would be tantamount to accepting the reality of climate change — at a time when the science was more tenuous than it is now. Developed countries were also concerned that accepting the need for adaptation amounted to an implicit assumption of responsibility, with the associated duty to compensate. At the same time, many developing countries were reluctant to discuss adaptation lest it derail developed country commitments to mitigation.[15] But as the science became clearer, and the failure of global efforts to reduce GHG emissions increasingly apparent, more attention shifted towards adaptation. At the Cancun negotiations in December 2010, the Parties to the UNFCCC established the Cancun Adaptation Framework,[16] in which Parties were requested to start making assessments of their vulnerability to climate change, plan adaptation actions, strengthen institutional capacities, build resilience and enhance their climate-related disaster risk reduction strategies.

By 2013, following the IPCC's Special Report Managing the Risks of Extreme Events and Disasters to Advance Climate Change Adaptation (SREX),[17] it had become apparent that many extreme weather and slow onset events were linked to a warming climate. Based on the best available science, the Parties established the Warsaw International Mechanism for Loss and Damage associated with Climate Change Impacts (the Mechanism),[18] under the Adaptation Framework. The Mechanism acknowledged that the loss and damage associated with climate change impacts cannot all be reduced by adaptation.[19] The Mechanism called on countries, amongst other things, to: undertake impact, vulnerability and adaptation assessments;[20] engage in climate resilient development,[21] enhance climate change disaster risk reduction;[22] and understand and cooperate on Climate Displaced Persons, migration and planned relocation at the national, regional and international levels.[23]

Even with the growing discussion around climate adaptation strategies over the past decade, there has been continued, if faltering, discussion of mitigation through control of greenhouse gas emissions. The most recent instalment, drafted in 2015 and signed in 2016, is the Paris Agreement,[24] which committed Parties to limit the increase in global average temperature to well below 2°C above pre-industrial levels, and pursue efforts to limit the temperature increase to 1.5°C.[25] For the first time, both developed and developing country Parties must prepare, communicate and implement successive voluntary nationally determined contributions (NDCs) that will be implemented through domestic mitigation measures. New NDCs must be communicated every five years and be informed by a Global Stocktake of emissions, starting in 2023.[26] Each successive NDC must represent a stronger target than the previous one, and developed countries are still expected to take the lead by undertaking economy-wide absolute emission reduction targets.

One of the important accountability mechanisms for Parties is the Enhanced Transparency Framework, which requires developed and developing countries to report every two years on progress towards meeting their emissions reduction targets. The information provided will be subject to a technical expert review, which will identify potential compliance issues and areas for improvement. A disappointing feature of the Paris Agreement is that it does not provide a basis for any liability or compensation for the impacts of climate change. However, a Task Force on Displacement was established to deal with the millions of people who will ultimately be displaced as a result of climate change.

Some believe that the Paris Agreement may be our final curtain call. Indeed, the United Nations Environment Programme's 2018 Emissions Gap Report[27] issues a warning that '[p]athways reflecting current NDCs imply global warming of about 3°C by 2100, with warming continuing afterwards. If the emissions gap is not closed by 2030, it is very plausible that the goal of a well-below 2°C temperature increase is also out of reach.'[28] Unfortunately, progress under the UNFCCC has moved at a snail's pace given the urgency of the project. Negotiations have threatened to collapse on many occasions,[29] and have involved astonishing brinkmanship among some of the key global leaders,[30] along with heroic and emotional appeals from developing country representatives. These past failures have shone a searching light on the weaknesses of multilateral negotiations, leading many to question the effectiveness of the legal enforcement mechanisms available through International Law.

It is clear that efforts to deal with the climate 'emergency' have been thwarted by domestic election cycles in fossil fuel-developed economies. Many politicians have either lacked the knowledge, or the political will, to lead a national discussion on the imperative to take action. Even worse, some politicians in the US, Australia and Canada, for example, have acted against the scientific consensus on climate change. Citizens have been encouraged to focus on the financial costs of carbon prices, as politicians chase the goal of winning government in short-term election cycles. Seldom is the current consensus on climate science clearly articulated and communicated to counter the deliberate undermining by the fossil fuel lobby and climate change skeptics. The potentially devastating impacts of climate change on economies and ecosystems are rarely discussed. Instead, the most significant climate change messages are lost as political sound bites resonate in the voting public's consciousness.[31]

Given the current state of affairs, some may regard the tenacity of the multilateral climate change negotiations as something of a miracle, especially in light of the tremendous changes the world has witnessed since 1992. Other pragmatists will know that walking away from the only negotiating platform for a global approach to climate change would leave nothing in its place. It is this acknowledgement that will keep the negotiations rolling on well into the future.

Endnotes

1. IPPC, *Global Warming of 1.5°C: An IPCC Special Report on the Impacts of Global Warming of 1.5°C above Pre-Industrial Levels and Related Global Greenhouse Gas Emission Pathways, in the Context of Strengthening the Global Response to the Threat of Climate Change, Sustainable Development, and Efforts to Eradicate Poverty*, ed. V. Masson-Delmotte et al., Geneva: World Meteorological Organization, 2018, https://www.ipcc.ch/sr15/

2. IPCC, *Climate Change and Land: An IPCC Special Report on Climate change, Desertification, Land Degradation, Sustainable Land Management, Food Security and Greenhouse Gas Fluxes in Terrestrial Ecosystems*, ed. H.-O. Pörtner et al., Geneva: IPCC, 2019, https://www.ipcc.ch/report/srccl/; See also 'Everyday Biodiversity' by Jeffrey R. Smith and Gretchen C. Daily in this volume.

3. IPPC, *Special Report on the Ocean and Cryosphere in a Changing Climate*, ed. H.-O. Pörtner et al., Geneva: IPCC, 2019, https://www.ipcc.ch/srocc/home/; See also 'Oceans 2020' by David M. Karl in this volume, and 'Ice' by Julian Dowdeswell in this volume

4. IPBES, *Global Assessment Report on Biodiversity and Ecosystem Services*, ed. M. Carneiro da Cunha, G. M. Mace and H. Mooney, Bonn, Germany: IPBES Secretariat, 2019, https://ipbes.net/global-assessment-report-biodiversity-ecosystem-services

5. Available at https://www.soas.ac.uk/cedep-demos/000_P514_IEL_K3736-Demo/treaties/media/1972%20Stockholm%201972%20-%20Declaration%20of%20the%20United%20Nations%20Conference%20on%20the%20Human%20Environment%20-%20UNEP.pdf

6. World Commission on Environment and Development, *Our Common Future*, Oxford: Oxford University Press, 1987, http://www.un-documents.net/our-common-future.pdf

7. Available at https://www.ipcc.ch/

8. Available at https://www.un.org/en/development/desa/population/migration/generalassembly/docs/globalcompact/A_CONF.151_26_Vol.I_Declaration.pdf

9. Available at http://sustainabledevelopment.un.org/content/documents/Agenda21.pdf

10. Available at http://www.cbd.int/convention/text/default.shtml

11. Available at https://digitallibrary.un.org/record/144461?ln=en

12. United Nations, *United Nations Framework Convention on Climate Change*, [n.p.]: UN, 1992 http://unfccc.int/resource/docs/convkp/conveng.pdf

13. See also 'Environmental Economics' by Don Fullerton in this volume.

14. Available at https://unfccc.int/resource/docs/convkp/kpeng.pdf

15. See E. L. F. Schipper, 'Conceptual History of Adaptation in the UNFCCC Process', in *The Earthscan Reader in Adaptation to Climate Change*, ed. by E. L. F. Schipper and I. Burton, London: Earthscan, 2009, 359–76 at 362.

16. The Cancun Adaptation Framework is contained within Articles 11–15 of the COP (Conference of the Parties) Report from the Cancun negotiations in 2010 (available at https://unfccc.int/resource/docs/2010/cop16/eng/07a01.pdf).

17. IPCC, *Managing the Risks of Extreme Events and Disasters to Advance Climate Change Adaptation: Special Report of the Intergovernmental Panel on Climate Change*, ed. C. B. Field et al., Cambridge, UK: Cambridge University Press, 2012, https://www.ipcc.ch/site/assets/uploads/2018/03/SREX_Full_Report-1.pdf

18. United Nations, *Report of the Conference on its Nineteenth Session held in Warsaw from 11 to 23 November 2013, Addendum, Part Two*, FCCC/CP/2013/10/Add.1, Decision 2/CP.19, http://unfccc.int/resource/docs/2013/cop19/eng/10a01.pdf

19. Ibid.

20. Ibid. Art. 14(b).

21. Ibid. Art. 14(e).

22. Ibid. Art. 14(e).

23. Ibid. Art. 14(f).

24. Available at https://unfccc.int/resource/docs/2015/cop21/eng/l09r01.pdf

25. Ibid. Art. 2(a).

26. Ibid. Art. 14(2).

27. United Nations, *The Emissions Gap Report 2018*, Nairobi: United Nations Environment Programme, 2018, http://wedocs.unep.org/bitstream/handle/20.500.11822/26895/EGR2018_FullReport_EN.pdf?sequence=1&isAllowed=y

28. Ibid. at xiv.

29. For example, the collapse of negotiations during the 2000 COP 6 at The Hague, the hastily convened COP 6.5 in Bonn 2001 and the failure of the 2009 Copenhagen COP 15 to adopt any decision on a legally binding agreement for the post-2012 era.

30. Such as the Premier Wen Jiabao declining President Obama's invitation to meet at Copenhagen until Obama, about to leave the COP, walked into a meeting of the BASIC countries and asked Premier Wen Jiabao whether he was ready to meet. It was this fortuitous meeting which resulted in the drafting in the final hours of the COP of the Copenhagen Accord.

31. On media coverage of climate change, see 'Media' by Candis Callison in this volume.

Weather

—

Neville Nicholls

I woke early on Sunday, 8 February 2009, in my house on the edge of Melbourne, the capital of the State of Victoria in southeast Australia. During the night, I had worked on a way to estimate the human health impact of extreme weather events, and I wanted to apply it to the heat wave that had hit southeast Australia the previous weekend, to see if the recently introduced heat wave alert system had helped save lives. I switched on my computer, and then the TV, to see the early reports of the bushfires that I knew would have occurred the previous day. For the past week, the Australian Bureau of Meteorology had been predicting that Saturday, 7 February, would be a day of very high temperatures and extreme bushfire risk. The Premier of Victoria was predicting the worst bushfire weather in the State's history.

With such dramatic predictions, and with an alert public and well-prepared emergency response crews, I expected substantial damage to properties, forests and farms, but few, if any, deaths. I was wrong. The first thing I saw on TV was the helicopter view of Marysville, a much-loved tourist town in the hills east of Melbourne. Or, more accurately, I saw a view of the charred and blackened remains of the town. I realized immediately that such destruction, caused by the combination of an unprecedented decade-long drought, record temperatures (46.4°C) and strong winds, would have led to many deaths. Indeed, more than 170 lives were lost across Victoria that day, despite the accurate predictions and warnings of an imminent public emergency.

          https://doi.org/10.11647/OBP.0193.18

As bad as the devastation was on February 7, Black Saturday, even more lives had been lost the previous weekend during an unprecedented heat wave when temperatures exceeded 43°C for three consecutive days. The problem I had been grappling with was to estimate, in near real time, the death toll from such heat waves. Traditionally, scientists studying the human impacts of heat waves would have to wait many months before the death toll could be calculated, relying on government bureaucracies to compile data from hospitals and nursing homes. But by comparing the number of obituaries in newspapers in the days after a heat wave with the numbers in more typical weeks, I could estimate the extra mortality caused by the heat wave. As I watched the TV footage of the destruction from what was already being called the Black Saturday bushfires, I started calculating the excess mortality of the previous week's heat wave. I hoped that the heat wave alert system would have meant far fewer lives had been lost than in heat waves in the past, but it quickly became apparent that, despite the forecasts and heat wave alerts, as many as 500 more people had died in southeast Australia over the previous weekend than would have typically been expected for that time of year.

The deadly 2009 Australian heat wave and the fires that accompanied the heat wave were just two of the many previously unprecedented weather extremes that have been observed in recent decades. These extremes show no sign of relenting, as witnessed by the massive drought-fuelled bush fires that consumed large areas of southeast Australia in 2020. Looking back over the past half-century, the impacts of climate change on extreme weather become increasingly clear. And as our understanding of extreme weather improves, so too does our forecasting ability, providing us with tools to increase resilience in an uncertain climate future. These tools may help us avoid some of the impacts of continued global warming, if politicians are unable or unwilling to act to slow this warming.

S cientists have been warning for decades that some extreme weather events would change in frequency or intensity as the world warmed due to the burning of fossil fuels. And indeed, over the past fifty years, this has proven to be the case.[1] High temperatures, including heat waves, have become more frequent. Globally, the number of warm days (days exceeding the ninetieth percentile of historical daily maximum temperatures) has doubled since 1970, while the number of cool nights (below the tenth percentile of historical daily minimum temperatures) has halved. The extreme weather associated with severe bushfire risk has also been increasing, while heavy rainfall events appear to be becoming more frequent.

In the absence of decisive political action to mitigate the global warming trend and its effects on extreme weather, bureaucrats and scientists around the world have begun developing and implementing new alert systems for extreme weather events, especially heat waves. This work has taken on an increased sense of urgency after the 2003 European heat wave that led to the death of as many as 70,000 people. By 2009, a heat event alert system had been established for Melbourne, based on the observation that mortality in the city increased substantially when the daily average temperature exceeded 30°C. An alert was initiated when forecasts by the Bureau of Meteorology indicated that this threshold would be crossed. Without this early warning system, even more people would likely have died in the 2009 Australian heat wave.

The development of early warning systems for extreme weather events has become possible in recent years because of improved weather forecasting capabilities. Today, in 2020, national weather services can forecast temperatures 5–6 days in advance more accurately than their predecessors in the early 1970s could forecast a single day in advance. These improvements have come from increased computing power, which has allowed ever more complex and realistic mathematical simulation of the atmosphere, and the increased availability of satellite observations to drive and refine model predictions. Thanks to these greatly improved weather forecasts, bureaucrats, politicians, the media, medical and emergency services and the public now have several days to implement strategies to minimize deaths from high temperatures and other extreme weather events. As global warming increases the frequency and intensity of heat waves, these alert systems can reduce some of the likely human cost of climate change.

Increased monitoring and forecasting skills allow us to observe the occurrence of a wide variety of weather extremes in addition to heat waves, and to predict their future trajectories. Extreme cold events also cause many deaths and illness in many parts of the world. Cold events have been decreasing in frequency and intensity in countries around the world, as another consequence of global warming. Nonetheless, cold events still occur, but these events can now be predicted days in advance, allowing us to reduce their health impacts. The combined effect of global warming, which reduces the frequency of cold events, and the improved forecasting of these events, will continue to reduce their deleterious effect on the human population.

Although the changing frequency of hot and cold extremes is very clear over the past fifty years, patterns in other extreme weather events are more difficult to identify. In some cases, this may reflect the absence of any real change in frequency of these events. But in other cases, it may be that changes in the way we observe particular weather phenomena are obscuring any real underlying change in their frequency or intensity. For instance, the increased availability of satellite observations since the 1970s has greatly improved our ability to detect tropical cyclones. Apparent increases in tropical cyclone activity may thus simply reflect better observations, rather than any real change in the frequency of cyclones.

Irrespective of any long-term trends, improved forecasting of tropical cyclones and other extreme weather, including storms, droughts and floods, has lessened their human impacts. Prior to the forecasting revolution of the last half-century, tropical cyclones were much more deadly. A single tropical cyclone in 1970, for instance, is believed to have caused more than 300,000 deaths in Bangladesh. Since that time, the mortality associated with tropical cyclones, even in countries with limited financial resources, has declined. For example, during Cyclone Fani in 2019, authorities in India and Bangladesh relied on improved monitoring and forecasting of the cyclone to move at least a million people out of the storm path, thereby drastically reducing the potential death toll. This reduction in the menace of tropical cyclones stems from improved satellite monitoring of these systems, and more accurate prediction of their trajectories, as well as the creation of improved alert systems and infrastructure to notify people of imminent danger and enable rapid evacuation. Looking to the future, cyclones, although still deadly, should not cause the enormous mortality seen half a century ago, even if their intensity increases on a warming planet.

Droughts are also changing as a result of global warming. Global warming has meant that droughts today are accompanied by higher temperatures than they would have been fifty years ago, and this trend towards warmer droughts will continue into the future. In some areas, there is also evidence that droughts have become more frequent (although this is not universal). Whether droughts in the future will be drier or longer-lasting will depend on the regional impacts of global warming on atmospheric circulation, which we cannot currently predict with confidence. It is even harder to predict how floods will change in the future, because of the complex factors at play. Over the past several decades, there has been an increase in intense precipitation events associated with global warming, and we can expect such changes to continue into the future. But the extent to which this increased rainfall will lead to flooding will depend, among other things, on alterations in the land surface (such as increasing road surfaces) and changes in riverbanks and drainage systems. Once again, our improved ability to monitor heavy rainfall events, using radar and satellites, has improved our ability to provide more timely forecasts of flooding. As these observing systems continue to evolve, they should help us avoid some of the greatest damage and threats posed by flooding, even if flooding events become more frequent.

Another significant improvement in our ability to cope with Earth's changing weather has been the development of seasonal climate forecasting. In some parts of the world, specifically those areas where the El Niño–Southern Oscillation (ENSO) dominates the inter-annual climate variability, the last fifty years have seen the development of scientific methods for forecasting seasonal droughts and extended heavy rain periods. Such seasonal forecasting of weather extremes was considered impossible in the 1970s, even by meteorologists and climate scientists. But by the mid-1980s, methods for seasonal forecasting of droughts had been developed, at least for some parts of the world. These seasonal forecasts, although lacking the skill achievable in near-term forecasting, provide hope that we might avoid some of the consequences of droughts and other inter-annual weather variability through adaptive crop and stock management, or the more timely provision of food relief. For example, if we can predict El Niño-related droughts across Pacific rim countries, we can ensure that drought relief arrives in a timely fashion, thereby avoiding famine in, say, the highlands of New Guinea, an area where droughts caused by the El Niño have led to severe famines in the past.

What about the smaller scale weather extremes, such as hailstorms and tornadoes? These short-term events are notoriously difficult to monitor, with historical records of such extremes relying heavily on subjective reports from observers. Such reports are, in turn, dependent on population density, amongst other factors. Thus, an increasing population in an area, for example, might lead to increased reports of hailstorms, even if the actual frequency is not changing. Disentangling such reporting biases from any real climate-forced change in these small-scale extremes is beyond our capabilities at present. We are thus left with little confidence in any apparent trends in these extremes, even over recent decades. Nevertheless, improved systems for detecting these small-scale extreme weather events, and our increased ability to issue and distribute short-term forecasts of their movement, have begun to allow populations to avoid some of the associated damages. Continued improvements in these monitoring and forecasting systems should help to further reduce the associated damage of short-lived extreme weather events, even if global warming increases their frequency or severity.

The past half-century has seen substantial changes in the frequency and intensity of some extreme weather events. But these fifty years have also seen advances in our ability to monitor and predict these extreme weather events (and others), thereby reducing the associated human impacts. In particular, meteorologists have vastly improved their ability to predict extreme hot and cold days, storms, bushfire weather and cyclones several days in advance. These improvements have led to the development of alert systems that have reduced the loss of life previously caused by such extremes. More can be done to improve these forecasts, and their public dissemination and use in communities. At the same time, seasonal climate forecasting has developed from a pie in the sky idea into a well-developed science, at least in some parts of the world and for some seasons. With this new tool, we can now predict some droughts and seasonal tropical cyclone activity, well in advance, providing opportunities for longer-term planning and disaster-reduction strategies.

The 2019–2020 Australian bushfire season provides a case in point. The economic damage of the fires will, no doubt, be huge, especially since the fires destroyed large areas where tourism is a major industry. But the human cost has been much lower than might have been the case if the fires had occurred twenty years ago. The long-range forecasts of a very severe fire season meant that the fires services were well prepared, and the shorter-range weather forecasts have helped them to fight the fires with greater effectiveness. Similarly, recent heat waves have led to fewer deaths than the 2009 heat wave, at least partly because of improved weather forecasts and heat wave alerts.

Looking forward, increasingly accurate weather forecasts, from days to months ahead, will allow us to reduce the damage — both economic and societal — that has been caused by weather extremes throughout human history, even as human actions lead to increases in the severity of some of these extremes. In the absence of concerted political action to slow the rate of global warming,[2] improved weather forecasting is perhaps the most important, immediate tool we have to offset some of the deleterious effects of human-caused climate change now and into the future.

Endnotes

1. J. Blunden, D. S. Arndt and G. Hartfield, 'State of the climate in 2017', *Bulletin of the American Meteorological Society*, 2018, 99, Si-S310, https://doi.org/10.1175/2018bamsstateoftheclimate.1; IPCC, 'Climate change 2013: The physical science basis', in *Contribution of Working Group 1 to the Fifth Assessment Report of the Intergovernmental Panel on Climate Change*, ed. T. F. Stocker, D. Qin, G.-K. Plattner, M. Tignor, S. K. Allen, J. Boschung, A. Nauels, Y. Xia, V. Bex and P. M. Midgley, Cambridge, UK: Cambridge University Press, 2013, 383–464, https://www.ipcc.ch/site/assets/uploads/2018/02/WG1AR5_all_final.pdf; IPCC, *Managing the Risks of Extreme Events and Disasters to Advance Climate Change Adaptation: Special Report of the Intergovernmental Panel on Climate Change*, ed. C. B. Field et al., Cambridge, UK: Cambridge University Press, 2012, https://www.ipcc.ch/site/assets/uploads/2018/03/SREX_Full_Report-1.pdf

2. See also 'Politics and Law' by Elizabeth May in this volume.

Knowing Earth

—

Sheila Jasanoff

The philosophers and scientists of the eighteenth-century Age of Enlightenment believed that increasing knowledge of how the world works would liberate humans from superstition, and that reason would follow learning. In its engagement with science, modern environmentalism seems to have turned back the clock on this view of Enlightenment. To be sure, environmentalism was born in partnership with science and technical expertise, but it achieved adulthood in controversy and matured in an era of growing skepticism and paralyzing uncertainty. Science and scientists, along with inventors and engineers, remain central to the environmental story, but few now believe that more science and better technology will enable humanity to become effective planetary stewards — as commanding captains of Spaceship Earth. Instead, the entanglement of science and environmental protection has been marked by advances and retreats, with science serving at times as a torch of illumination, and at others as a lightning rod for controversy. Unquestionably, advances in science have allowed us to know Earth, and our place within it, far differently than on the first Earth Day five decades ago. Changing knowledge, however, has not brought greater mastery, as optimists of that earlier era might have hoped. Instead, scientific knowledge today confronts humankind with the challenge of assuming greater responsibility for an Earth whose complex dynamics elude full understanding, and whose very capacity to sustain human life is seen by many as gravely threatened.

    https://doi.org/10.11647/OBP.0193.19

In the rapidly industrializing nineteenth century, the urge to protect nature arose in an acutely emotional register; a sense of irreparable loss as nature's tranquil beauty was ravaged by the smoke, filth and noise of the machine age. Only gradually did people learn that producing goods on mass scales not only did violence to pristine landscapes, but also harmed the health and wellbeing of all living things and the ecological and biophysical systems that sustain them. Biologist and natural historian Rachel Carson is widely credited with sounding an alarm that could not go unheeded. *Silent Spring*, her 1962 broadside against chemical pesticides,[1] helped ignite a social movement, calling attention to the stealthy, lethal pathways by which human-made toxins indiscriminately accumulate in organisms far beyond the intended targets of industrial 'biocide'. Benign DDT (dichlorodiphenyltrichloroethane), once known as a potent weapon against malaria and typhus, turned in Carson's telling into a symbol of technological over-reach, decimating bird populations, causing cancer in test animals (though not demonstrably in humans) and coursing in mothers' milk. Chemicals joined nuclear radiation as invisible bearers of harm whose unforeseen and unpredictable impacts potentially outweighed their acclaimed economic and health benefits. These dangers could neither be sensed nor entirely guarded against; they made us all reluctant denizens of what the German sociologist Ulrich Beck called the 'risk society'.[2]

Rising awareness of chemical and nuclear risks in the 1960s proved to be a boon for the environmental sciences. Indeed, one can see the 1970s as a decade of achievement in the institutionalization of scientific studies of the environment as well as in environmental law and policy. The US federal government led the way in research with the formation of the National Institute of Environmental Health Sciences (1969) and the National Toxicology Program (1978), as well as expansions in the scientific capabilities of regulatory bodies such as the newly formed Environmental Protection Agency (EPA). Universities followed suit, creating new departments and programs to study the environment in all its dynamic variety. In the private sector, the Ford Foundation sponsored opportunities for lawyers and scientists to collaborate for environmental protection through grants to influential organizations such as the Natural Resources Defense Council and the Environmental Defense Fund. Companies, too, recognized a need to develop new forms of expertise to meet the burdens of information production and risk assessment created by the changing landscape of environmental law.

Despite a promising start, the end of the 1970s brought backlash against environmental expertise, especially in the United States. Alarmed by the increasingly close partnership between science and governmental regulation, industry representatives launched systematic attacks on the quality of regulatory science, questioning both its validity and the integrity of its practitioners.[3] Much of regulatory knowledge, opponents insisted, was predictive, speculative and drawn from questionable or indirect sources, such as animal tests to determine the likelihood of human cancer, or climate models to predict the rise in global temperatures. Critics charged that much of this science was not peer-reviewed or published in reputable journals. Without well-established paradigms to guide environmental research and risk assessment, and with almost infinite entry points for questioning the methods and assumptions underlying science-based policy, the EPA proved particularly vulnerable to the politics of the moment. Well supported by the Clinton and Obama administrations, but aggressively undermined during the presidencies of both George W. Bush Jr. and Donald Trump, the EPA lost its global leadership position in the delivery of reasoned, science-based environmental protection. One telling indicator of the EPA's declining influence and regulatory muscle was its persistent failure to implement the Toxic Substances Control Act, a federal law that most directly responded to the threats laid bare by *Silent Spring*.[4]

Despite much controversy, environmental science continued to make large strides, especially in its power to detect and explain planetary phenomena. A notable success story of the late twentieth century was the detection of the ozone hole in the mid-1980s, which offers perhaps the best example to date of the rapid and effective integration of science and policy.[5] In a 1974 article in *Nature*, future Nobel laureates F. Sherwood Rowland and Mario J. Molina published troubling findings about the likely effects of chlorofluorocarbons (CFCs) — widely used in refrigerators, air conditioners and spray cans — on the stratospheric ozone layer shielding Earth from harmful ultraviolet radiation. Subsequent observations confirmed their disturbing hypothesis and, in 1987, leading industrial nations signed the Montreal Protocol, an international pact to ban and phase out CFCs and other ozone-depleting chemicals.[6]

Scientific consensus in this case pulled global policy in its wake. By 2015, all members of the United Nations were on board with the agreement to phase out the harmful substances, and the 'hole' in the ozone layer is now showing signs of gradual recovery.

Successful as it was, the CFC story also carried warnings about the limits of scientific knowledge when confronting problems of global scale and huge economic consequence. For at least a decade before the Montreal Protocol, uncertainty and indeterminacy served as rallying points for industry opposition to a CFC phase-out, and the policy consensus proved anything but straightforward to implement. As late as 1986, DuPont, the largest producer of CFCs, still led industry efforts to discredit the science advanced by Rowland and Molina. The company changed its tune only after developing new profitable compounds in an emerging market for CFC substitutes. The Montreal Protocol itself was negotiated among a relatively small group of producer nations, and special provisions were needed to draw in developing countries that were, if anything, more dependent on cheap refrigerants.[7] Even then, production of fluorocarbons never completely ended, and periodic violations of the letter and spirit of the ozone accords continued well into the new century. The Montreal Protocol nonetheless stands as a significant achievement for enlightened environmental policy. A risk was identified, and its cause eliminated. The treaty garnered formal support from all of the world's nations, and potentially catastrophic consequences were averted because politics followed where the science pointed.

The story of climate science traces a less triumphalist narrative line. The science in this case focused on the effects of carbon-containing greenhouse gases (GHGs) on Earth's average surface temperature. The idea was not new. Around 1896, the Swedish physical chemist Svante Arrhenius posited for the first time that GHGs released by human activity would cause the Earth to warm. Simulating the internal dynamics of a greenhouse, GHG molecules would trap radiant heat from the Earth's surface, absorbing it into the atmosphere and directing more heat back toward the planet, thus making temperatures rise. Since those early speculations, more than a century ago, many observations have converged to establish the truth of Arrhenius' hypothesis as solidly as any major finding in earth and planetary sciences.

If it takes a village to ensure the well-being of young children, then it is hardly surprising that it took a massive, collective effort to establish the scientific facts of climate change. Since 1988, that work has been led by the Intergovernmental Panel on Climate Change (IPCC), a body created by the UN Environment Program and the World Meteorological Organization to assess the mountains of data on Earth's changing biophysical systems, and to clarify the nature and severity of those changes in relation to human well-being. Divided into three working groups — on science, impacts and policy — the IPCC has always insisted on its political neutrality. Its work, the IPCC repeatedly asserts, is policy-relevant but not policy-prescriptive. Yet, it soon became apparent that policy neutrality could hardly remain a realistic option if the IPCC's claims were to be taken at face value. Since its first Assessment Report (AR) in 1992, the IPCC has issued a total of five ARs (a sixth is in the offing), and many additional special reports on specific effects and assessment methods. The basic conclusion that human activities are heating the planet has hardened with each report, while warnings have become more urgent that Earth is headed toward a point of no return, with melting ice caps, unpredictable sea level rise and extreme weather patterns endangering billions of vulnerable lives around the globe. These dire scenarios motivate the evangelical fervor of today's climate activism, in which many scientists engage along with lay citizens.

Unlike the case of ozone depletion, political action on climate change failed to keep up with the urgency of the scientific predictions. In 2015, the nations participating in the UN Framework Convention on Climate Change (UNFCCC) agreed to implement national reductions in CO_2 emissions that would hold global temperature rise to 2°C or, better still, 1.5°C (the Paris Agreement, signed in 2016).[8] Just five years later, the lower target seemed almost unattainable, and none of the highest emitting countries appeared on track to meet their Paris obligations. Most shockingly, in the face of widespread criticism, US President Donald Trump withdrew his country from the Paris Agreement soon after his election, arguing that meeting the treaty obligations would harm the American economy, placing businesses and workers at an unfair disadvantage. That economic argument continues to sway an electorate that has become less skeptical about the fact of anthropogenic climate change and yet remains reluctant to make the economic sacrifices and lifestyle changes needed to significantly curb GHG emissions.

If genuine reductions in GHG emissions are to be achieved, humanity will need to harness not only science and technology, but also its collective moral will. The astronomer and gifted science popularizer Carl Sagan offered a foretaste of that thought in his 1994 book, *The Pale Blue Dot*.[9] The floating images of Planet Earth brought home to Sagan the smallness and isolation of human existence. There is no sign in the vastness of space that humanity's salvation will come from anywhere else other than Earth and its human inhabitants. Hence, this 'thin film of life on an obscure and solitary lump of rock and metal' has to care for itself. As Sagan observes, 'To me, it underscores our responsibility to deal more kindly with one another, and to preserve and cherish the pale blue dot, the only home we've ever known'.[10]

E ven as climate action remains unsteady and contested, there are some indications that the industrial world has begun to accept another truth that would have seemed self-evident to our premodern ancestors, and to those outside the Judeo-Christian tradition: human beings do not stand above or apart from nature, but are instead deeply embedded within it. Environmental sciences have increasingly shown how collective human activities are altering planetary dynamics to the potential detriment of our own and other species. The very idea of the Anthropocene, a new geologic age marked by the human imprint on the planet, points to the inseparability of nature and culture.[11] Outside the contentious debate on fossil fuel use, these insights are fostering new forms of environmental responsibility and political engagement. The worldwide movement against single-use plastic products, for example, signals a desire to clean the oceans of debris that threatens marine life and ecosystems. Other large collective actions range from decreased meat consumption and nationwide tree-planting campaigns to youth movements challenging their elders' perceived indifference toward the disastrous implications of climate change. Scientific insights are woven into these movements for change. The young climate activist Greta Thunberg, the living face on movements like Fridays for Future, draws her moral conviction squarely from science as she sees it. But the connections between science and social movements are not direct; the influence of science on Thunberg and millions of others around the world are tied less to methodologically rigorous demonstrations than to the perception that living unsustainably on Earth is no longer ethically tenable.

What do these developments mean for environmental science and politics in the twenty-first century? Clearly, there is no question that more scientific knowledge is needed. If anything, a growing awareness of the interconnectedness of Earth's living and non-living systems heralds a new age of discovery across the entire spectrum of the environmental sciences. The rise of intersecting and hybrid fields, such as biogeochemistry and sustainability science, attest to scientists' recognition that new understanding will have to be sought at the intersections of older fields and outside the boundaries of traditional disciplines. The enormous power of computing and data science have opened up new possibilities for modeling Earth's future on scales and at levels of detail that were not imaginable fifty years ago. Inspired individual insight will still retain a place, as it always has in science, but the scientific future belongs increasingly to centers and collectives capable of drawing together knowledge from multiple fields.

The events of the past half-century have taught us that mere gains in scientific understanding will not translate into wise policies for the human future. To enable that translation, we will need to harness all we have learned about making knowledge actionable and persuasive. This means, in the first instance, understanding that the environmental sciences cannot exist purely in the realm of impartial facts, cordoned off from political discourse and moral imperatives. Scientists must learn to see that describing the world in new terms demands that we change our ways of living in the world, to accommodate both what science tells us *and* what it is unable to foresee. The politics of environmental science in the next half-century will have to build on the understanding that science and planetary stewardship are co-produced. Inevitably, the politics of the Anthropocene will also have to be a politics of precaution.

Endnotes

1. R. Carson, *Silent Spring*, New York: Houghton Mifflin, 1962.

2. U. Beck, *Risk Society: Towards a New Modernity*, London: Sage Publications, 1992.

3. S. Jasanoff, *The Fifth Branch: Science Advisers as Policymakers*, Cambridge, MA: Harvard University Press, 1990.

4. See also 'The Global Chemical Experiment' by Elsie Sunderland and Charlotte C. Wagner in this volume.

5. See also 'Air' by Jon Abbatt in this volume.

6. Available at https://ozone.unep.org/treaties/montreal-protocol-substances-deplete-ozone-layer/text

7. See also 'Politics and Law' by Elizabeth May in this volume.

8. Available at https://unfccc.int/process-and-meetings/the-paris-agreement/the-paris-agreement

9. C. Sagan, *The Pale Blue Dot: A Vision of the Human Future in Space*, New York: Random House, 1994.

10. Ibid., at 3 and 7.

11. J. Purdy, *After Nature: A Politics for the Anthropocene*, Cambridge, MA: Harvard University Press, 2015.

Fish

—

U. Rashid Sumaila and Daniel Pauly

Humans have relied on coastal fish and other marine life for millennia. The first documented cases of human use of marine resources, include 165,000 year old abalone shells in a South African cave, 125,000 year old middens along the coast of Eritrea with shells of a now extinct giant clam species, and sophisticated harpoons from the Congo dating back 90,000 years.[1] For much of our history, our impact on fish species was small, and the supply of fish must have seemed inexhaustible.

In our early fishing days, energy was provided by the muscles of fishers, and later by the wind. Even with sophisticated sailing vessels, the energy density that could be used for fishing was limited, and ultimately dependent on sunlight, which grew the food that fishers ate and fueled the winds that powered sailing vessels. This changed radically in the 1880s, when steam-driven trawlers began to be deployed around the British Isles, the home of the Industrial Revolution.[2] This development heralded the replacement of muscle and wind power by fossil fuel energy accumulated over millions of years in the form of coal.

The new trawlers — although inefficient by today's standards — made short thrift of the accumulated biomass of large fish around the British Isles, and soon had to move offshore, into the open North Sea, and later into the open North Atlantic, to maintain their catch rates. As other industrialized countries adopted the wonderful British innovation,

          https://doi.org/10.11647/OBP.0193.20

the highly concentrated energy of fossil fuels soon began to overwhelm the productive capacity of natural fisheries, much as motorized chainsaws would eventually overcome the productive capacity of many forests. It was during these final years of the nineteenth century, with the ascent of industrial fishing, that the world's marine fish populations first began to feel the mounting pressure of humanity.

The resulting declines in fish populations prompted the founding, in 1892, of the International Council for the Exploration of the Sea, the first international organization devoted to studying fisheries and fish populations, which was headquartered in Copenhagen, Denmark. Around the same time, declining fish stocks also prompted the emergence of fisheries science as a modern scientific discipline in the early twentieth century, first in Europe, the USA and Canada, then spreading worldwide, all the way to Japan. However, despite tentative efforts by the short-lived League of Nations between the two World Wars, it was only after the Second World War, with the establishment of the United Nations' Food and Agricultural Organization (FAO), that the systematic collection and analysis of global fisheries data began. Importantly, this was a time when most countries in Africa and large part of Asia still existed as European colonies, subjected to exploitative disruptions of their cultures and economies, including their national fisheries.

The FAO was tasked with providing scientific information on global food production, with the ultimate goal of eliminating hunger worldwide. To lay the foundation for this important goal, the organization began compiling global production data for key food commodities, including, marine fish. Thus began the FAO global fisheries catch statistics, issued annually since 1950. Until recently, the FAO statistics were the only source of global fish catch information, and it was these data that guided the development of fisheries management over the latter half of the twentieth century.

Unfortunately, the FAO data suffered a number of significant limitations, and these became apparent only gradually. For one thing, many catch data did not include important components, such as the contribution of small-scale and recreational fishers, as well as millions of tonnes of discarded bycatch. These catches were not considered, given the focus on industrial fishing and tradable commodities, as opposed to the employment and food security benefits of artisanal and subsistence fisheries.[3] Moreover, no attempt was made to account for illegal, unreported and unregulated (IUU) fishing activities and catches, even when the former were blatant and the latter estimable.

These omissions resulted in a significant under-reporting of the global marine fish catch (by as much as 50%)[4] with the result that global fisheries management was based on incomplete information about the true fishing pressure. Moreover, the considerable catch of small-scale sectors (artisanal, subsistence and recreational) was systematically underestimated, thus providing a perverse 'justification' for the continued marginalization and neglect of small-scale fishers.

The FAO's early focus on commercial fishing was further entrenched by the establishment of national fisheries management institutions set up primarily to serve the interest of commercial fishing enterprises. It was all about how much could be caught and marketed. This meant that Departments of Fisheries were placed under the Ministry of Agriculture in many countries, and in the United State, under the Department of Commerce. During the post-war decades, conservation and the ecosystem effects of fishing were never explicitly part of fisheries management concerns. Also, small-scale fisheries were further marginalized, with few if any studies on, for example, subsistence fisheries, in spite of their importance to the food security of numerous small island developing states.[5]

The narrow focus on commercial fish catch during the post-war years posed a significant challenge for effective management of the world's fisheries. Without comprehensive knowledge of the total marine biomass removed from the ecosystem, management was conducted in the dark, resulting in highly variable and unsustainable catch levels. At the same time, global demand for seafood began to increase sharply with rising world population and income after the Second World War, while the development of ever more sophisticated fishing technology led to more efficient fishing.

Technological innovations and changing fishing practices led to rapidly increasing annual catches throughout the 1960s, and at the time of the first Earth Day, in 1970, there was little to suggest that the world's fisheries were in danger. In fact, global catches continued to increase year after year, if only through the continuous expansion of the fishing grounds exploited by distant-water fleets, mainly toward southern latitudes and into deeper waters. It wouldn't be long, however, before signs of trouble began to appear.

The first collapses occurred in the California sardine fishery (of 'Cannery Row' fame) in the early 1950s, and the Peruvian anchoveta fishery in 1972, then the largest fishery in in the world. Both these collapses could be attributed to natural oceanographic processes, including, strong El Niño conditions in the case of Peru, which warmed surface waters and reduced the supply of nutrients for plankton. In contrast, the collapse of the Norwegian spring spawning herring in the 1960s was the first to be linked directly to overfishing. The most blatant case of overfishing, combined with the active suppression of the voices that warned about an impending collapse, would come about twenty years later, with the demise of the Northern cod fishery of Canada. Within just twenty to thirty years, giant trawler vessels reduced the cod population to less than one percent of its historical abundances, which had consistently yielded annual catches from small vessels and traps ranging between 100,000 and 200,000 tonnes over the previous five centuries.[6] And yet, when the fishery was closed in 1992 resulting in the loss of 40,000 jobs, Canada's fishery managers blamed the seal predators and a cold winter.

Other collapses, large and small, began to accumulate in various parts of the world, and, in 1996, the global catch trend began to decline for the first time since the post-war years. In other words, 1996 was the year of peak global catch, and catches have been dropping ever since, despite increasing fishing effort. Other impacts of fishing on fish populations also became apparent. Most notably, fishers began to see a change in the kinds of fish being caught, as large predatory fish on top of marine food web (e.g., cod, swordfish and tuna) became increasingly scarce, while smaller forage fish, such as herrings, sardines and anchovies, began to make up an increasing proportion of total catches. This 'fishing down the food web'[7] had a significant impact on ecosystem functioning, and also on ocean sea floor habitats, which became increasingly affected by the use of trawls and other bottom-impacting gear. This led to a decrease in the resilience of both fish stocks and marine ecosystems.

The 'fishing down' concept and the development of ecosystem-wide thinking began to broaden fisheries management to include the ecosystem effects of fishing in general.

This drew attention not just to the quantity of fish taken by fisheries, but the total removal of marine living organisms from ecosystems, and the impacts of various fishing gear on their habitats and biodiversity. And as fisheries and marine science began moving away from a narrow focus on the maximum catch that could be taken from marine ecosystems, fisheries economics, management, governance and policy were also having their moments of reckoning. With more and more stocks collapsing, the world could not continue pretending that the oceans and the great global sea fisheries were inexhaustible. This led to the introduction of various management measures that have continued to evolve with time.

Even before Garrett Hardin's famous paper, 'The tragedy of the commons' was published in *Science* in 1968, economists had argued that the problem of overfishing stems mainly from the common property nature of wild fish stocks, coupled with the absence of effective access or property rights to the fisheries resources.[8] In such an unregulated 'common pool', fishers are presented with a set of perverse incentives to overexploit the resource and dissipate its economic returns. By the late 1950s, this realization led managers, with support from fisheries scientists and economists, to begin putting in place restrictions to fishing. The motivation for this approach was to remove incentives to overexploitation with gear restrictions, and importantly, with global catch limits (total allowable catches, or TACs) enforced by monitoring and surveillance systems.

During the 1970s, TACs became increasingly adopted around the world, but it did not take long before it became clear that this approach, on its own, was insufficient to protect fish stocks. For one thing, appropriate TACs, based on sound ecological understanding of fisheries populations, were difficult to set, and even more difficult to enforce. Second, even if TACs could be properly determined and enforced, fishers would compete for catch shares, inevitably resulting in the build-up of excess fleet capacity and economic waste. Fisheries economists' termed this state of affairs 'regulated open access'.

Disappointed with regulated open access, management approaches evolved to include limited entry schemes in which TACs were combined with restrictions on the number of vessels that could participate in the fishery. Often, this approach was accompanied by a vessel buyback program,[9] in cases where fishing capacity was already in excess of sustainable levels, as was the case for the Canadian Pacific salmon fishery. Once again, however, fisheries managers quickly became disillusioned by limited entry schemes, as these proved to have a limited effect on over-fishing, and fishing capacity continued to increase, resulting in both economic waste and ineffective controls over total catch.

Starting in the late 1980s and early 1990s, economists frustrated with the lack of success with existing management schemes came up with a new approach designed to better align fishers' incentives with the best interests of society. The foundational argument for this new approach was the need to create *de facto* access and property rights, whether at the community, public or private levels. A number of alternative ways to creating such rights were suggested, including individual transferable quotas (ITQs), fishers' cooperatives, community-based fisheries management, and various combinations of these approaches. But it soon became clear that there was no 'best' alternative for all cases, and that a fishery by fishery approach would be required. For example, ITQs can face monitoring problems, associated with 'quota busting' (illegally catching more than one's quota) and 'high grading' (catching one's quota, then discarding it to catch larger fish, which fetch a better price). And in almost all fisheries, monitoring, control and surveillance are still inadequate, particularly in the case of many developing country inshore fisheries, leading to a breakdown of ITQ schemes. In addition, there are social concerns with ITQs where a few individuals or groups get the right to exploit public resources, in most cases free of charge.

Based on the challenges of previous quota-based systems, the past decade has seen an increased movement towards outright protection of some global fish stocks in at least part of their natural range. In particular, there has been a growing focus on the establishment of marine protected areas (MPAs), where fishing activities are tightly regulated and more rigorously enforced. Small and large MPAs, ranging from fully protected marine reserves to partially protected ocean areas have been introduced in many maritime countries to protect marine biodiversity more broadly, and to serve as a buffer against management errors and scientific uncertainties. Several international frameworks, including the 2011 Aichi Targets of the Convention on Biological Diversity, and the United Nations Sustainable

Development Goals have set a target of creating MPAs covering at least 10% of the ocean surface area by 2020. Unfortunately, we have fallen significantly short of this target, with only 2% of the world's MPAs currently in strongly or fully protected areas.[10]

Over the past half-century since the first Earth Day, the nature of global fisheries has changed significantly. These five decades have seen large decreases in the abundance of fish almost everywhere, along with broader effects on the tropic composition that ripple throughout marine food webs. At the same time, fisheries science and management have evolved from narrow beginnings in the post-war years, to a suite of broader approaches designed to match the realities of overexploitation of fish stocks and the marine ecosystems in which they are embedded. Science and management are evolving from focusing on maximizing commercial catch to broader conservation and sustainable development goals that are inclusive of different ocean stakeholders.

Although things look dire, there is a path forward to a better future. The best fisheries science available shows that the decades-long global decline in fish catches could be reversed if the world's maritime countries reduced the fishing effort in their exclusive economic zones. This, together with the removal of harmful subsidies,[11] the elimination of illegal fishing, a greater emphasis on future benefits and the closure of the high seas to fishing, would allow the fish to rebuild their abundance, and allow for higher catches than at present. The problem is that either the politicians do not accept the results of fisheries science (as is sometimes also the case for climate science), and/or are unable to stand up to the industrial fisheries lobby, to which they have largely ceded the exploitation and quasi-ownership of public marine resources. Unless things change, largely uncontrolled industrial fishing is likely to continue until the bitter end — whatever that may be. Add marine pollution to this (including plastic) and the multiple ocean stressors generated by climate change (sea surface temperature rise, ocean acidification and deoxygenation),[12] and the future might look rather bleak. We desperately hope that the world takes suitable action on multiple fronts to chart a different path forward.

Endnotes

1. J. E. Yellen, A. S. Brooks, E. Cornelissen, M. J. Mehlman and K. Stewart, 'A middle stone-age worked bone industry from Katanda, Upper Semliki Valley, Zaire', *Science*, 1995, 268, 553–56, https://doi.org/10.1126/science.7725100

2. C. Roberts, *The Unnatural History of the Sea*, Washington, DC: Island Press, 2010.

3. L. C. Teh and U. R. Sumaila, 'Contribution of marine fisheries to worldwide employment', *Fish and Fisheries*, 2013, 14, 77–88, https://doi.org/10.1111/j.1467-2979.2011.00450.x

4. D. Pauly and D. Zeller, 'Catch reconstructions reveal that global marine fisheries catches are higher than reported and declining', *Nature Communications*, 2016, 7, https://doi.org/10.1038/ncomms10244

5. D. Zeller, S. Harper, K. Zylich and D. Pauly, 'Synthesis of under-reported small-scale fisheries catch in Pacific-island waters', *Coral Reefs*, 2015, 34, 25–39, https://doi.org/10.1007/s00338-014-1219-1.

6. C. Walters and J. J. Maguire, 'Lessons for stock assessment from the northern cod collapse', *Reviews in Fish Biology and Fisheries*, 1996, 6, 125–37, https://doi.org/10.1007/BF00182340

7. D. Pauly, V. Christensen, J. Dalsgaard, R. Froese and F. C. Torres, 'Fishing down marine food webs', *Science*, 1998, 279, 860–63, https://doi.org/10.1126/science.279.5352.860

8. G. Hardin, 'The tragedy of the commons', *Science*, 1968, 162, 1243–48, https://doi.org/10.1126/science.162.3859.1243

9. C. W. Clark, G. Munro and U. R. Sumaila, 'Subsidies, buybacks, and sustainable fisheries', *Journal of Environmental Economics and Management*, 2005, 50, 47–58, https://doi.org/10.1016/j.jeem.2004.11.002

10. E. Sala, J. Lubchenco, K. Grorud-Colvert, C. Novelli, C. Roberts and U. R. Sumaila, 'Assessing real progress towards effective ocean protection', *Marine Policy*, 2018, 91, 11–13, https://www.sciencedirect.com/science/article/pii/S0308597X17307686, https://doi.org/10.1016/j.marpol.2018.02.004

11. U. R Sumaila., N. Ebrahim, A. Schuhbauer, D. Skerritt, Y. Li, H. S. Kim, T.G. Mallory, V. W. L. Lam and D. Pauly, 'Updated estimates and analysis of global fisheries subsidies', *Marine Policy*, 2019, 109, 103695, https://doi.org/10.1016/j.dib.2019.104706

12. See also 'Oceans 2020' by David M. Karl in this volume.

The Global Chemical Experiment

—

Elsie Sunderland and Charlotte C. Wagner

Since antiquity, humans have mined toxic elements such as mercury (Hg) and lead (Pb) from Earth's crust. The ancient Romans used Pb for their plumbing and kitchenware, while early Chinese and Egyptian civilizations colored their clothing and artwork with the brilliant red pigment of the Hg-enriched mineral, cinnabar. Paracelsus, the sixteenth-century physician and forefather of the field of toxicology, was an advocate for treating syphilis with mercury vapor rooms. Several hundred years later, nineteenth-century hat makers (the so-called mad-hatters) in Europe and North America inhaled acutely toxic levels of elemental mercury during the felting process. Today, fifty years after the first Earth Day in 1970, exposure to chemical toxicants has reached a planetary scale — a global chemical experiment in which we are all unwitting participants.

As the human population has grown exponentially over the past century, there has been an increased reliance on Pb, copper (Cu), zinc (Zn), and other heavy metals as 'technological nutrients' fueling modern industry.[1] In addition, almost 100,000 synthetic organic chemicals have been developed for everyday domestic and commercial use, in pursuit of 'better living through chemistry',[2] compliments of DowDuPont (the merger company of the Dow Chemical Company and DuPont, commonly still referred to as DuPont) and other major conglomerates. Synthetic fertilizers and pesticides have enabled the Green Revolution,[3] thereby avoiding the Malthusian population time bomb. At the same time, we are now undergoing a materials revolution, where the objects of the past, designed

 https://doi.org/10.11647/OBP.0193.21

for a single purpose (think a desk, or book), are being replaced by 'smart materials' with multi-functionality. This revolution has led to spectacular combinations of the elements from across the periodic table into modern materials and gadgets, with capabilities well beyond what was imaginable even a generation ago. As a result, we are now faced with tens of thousands of chemicals used in our homes and in our everyday products. For the vast majority of these chemicals, we have only a limited understanding of how they will behave once released into the environment.

A major problem is that many synthetic chemicals are persistent in nature, meaning they do not readily break down following release into air, water or soils. This persistence allows the chemicals to accumulate in the environment, and to be transported long distances from their original sources if they are released to air or water. Heavy metals occur naturally on Earth, but their abundance in environments where they are likely to accumulate in organisms has increased dramatically due to human activities. For example, cumulative anthropogenic releases of Hg over the past 500 years have been fifteen times higher than those from natural sources such as volcanoes. Once released to the environment, chemical contaminants can be transformed into forms that are readily taken up by living organisms, resulting in a process of biomagnification up the food chain, with increasing concentrations encountered in each successive step from prey to predator. Apex predators, including humans, sit at the top of the trophic pyramid, and are thus most vulnerable to persistent and bioaccumulative chemicals. As a result, the blood of nearly every mammal on the planet, from humans in Madagascar to polar bears in the Arctic, now contains a cocktail of global toxicants such as chlorinated, brominated and fluorinated synthetic organic compounds, Hg and Pb, among other substances.

By the time of the first Earth Day in 1970, chemical contamination of air, water and soil was well established. High profile cases, including dichlorodiphenyltrichloroethane (DDT) and other pesticides, along with wide-spread industrial pollution of various lakes and waters, prompted the formation of the US Environmental Protection Agency (US EPA) in 1970, and milestone legislations such as the Clean Water Act[4] and the Clean Air Act (US).[5] Yet, despite growing awareness of the threat of pesticides and other toxic chemicals, significant human impacts on natural ecosystems and human health continued. The deadliest example of such impact was the gas leak and explosion that occurred in December 1984 at the Union Carbide pesticide plant in Bhopal, India. Over 500,000 people were exposed to methyl isocyanate gas, and more than 2,000 eventually died from acute symptoms. The ground hugging deadly fog was a powerful image of the toxic effect of pesticides, but even more dire was the contamination of soils and groundwater, which led to an estimated 15,000–20,000 premature deaths in the subsequent two decades, painting a dark picture of the long-term effects of chemical exposures.

Awareness of the health risks associated with exposure to environmental contaminants among the general public has typically depended on acute exposures or mass poisonings resulting in visible health effects or death. In their landmark 2006 review paper in *The Lancet* medical journal, Drs. Philippe Grandjean and Philip Landrigan eloquently describe a predictable path for society's understanding of the health costs of industrial chemical exposures.[6] Acute poisoning events spark public interest and support for expensive research on a few select compounds. After several decades of research, a weight of evidence is established showing the 'silent pandemic' of health effects associated with chronic, low level exposures to globally-ubiquitous pollutants. Perhaps the best-known examples of this are Pb and Hg.

More than a century ago, poisoning of young children by peeling flakes of Pb-based paints was documented in Australia, and this was followed by repeated cases in Europe and North America. During the 1970s and 1980s, while we were happily combusting and releasing huge quantities of Pb as tetraethyl lead in gasoline, Bruce Lamphear, now at Simon Fraser University in Canada, conducted pioneering research linking children's blood Pb levels to IQ deficits.[7] Based on Lamphear's work and that of others, the level of blood lead considered 'safe' by the US Center for Disease Control (CDC) dropped from 60 mg/dL in 1960 to <5 mg/dL in 2012. Researchers now believe that all levels of Pb exposure may be associated with neurodevelopmental deficits, and problems of lead contamination still remain. In 2014, for example, high Pb levels were reported in the drinking water of Flint, Michigan due to a shift in their water source to the Detroit River, and the leaching of Pb out of acidic supply pipes.

Mercury contamination, like Pb, also gained significant public attention during the second half of the twentieth century, in response to several high-profile public health disasters. In the late 1950s, industrial discharges of an organic form of mercury into Minamata Bay, Japan led to the now infamous cases of Minamata disease, an acute form of Hg poisoning that results in tremors, impaired vision, memory loss, hair loss, birth defects and death. In the 1970s, an international food aid shipment to Iraq included grain seeds coated with an organomercury compound. The starving population consumed the grain in bread rather than planting the seeds, as intended, leading to severe mercury poisoning across the population. Today, many members of the public and the media incorrectly associate relatively low-level methylmercury exposure through consumption of fish with risks of Minamata disease. However, like Pb, there appears to be no lower threshold for chronic low-level methylmercury exposure and neurodevelopmental delays.

An emerging class of fluorinated organic compounds, poly- and perfluoroalkyl substances (PFAS), are this generation's organochlorine pesticides. These compounds have been used since the 1960s, mainly for their surfactant properties in various products ranging from Teflon pans, outdoor gear and food packaging, to aqueous fire-fighting foams. They are known as 'Forever-Chemicals' because they contain a fluorine (F)–carbon (C) bond that is not known to degrade under natural conditions. Like Pb and Hg, awareness of the health risks associated with these chemicals was ignited by the public outcry of communities located next to manufacturing facilities with contaminated water, soils and food. High levels of exposure to PFAS have been linked to wide range of health issues, from cancer to thyroid disease.[8] At low exposure levels, these compounds have been associated with the most potent immune-toxic response ever documented for a synthetic chemical present in the environment. In response to these acute health risks, the US EPA lowered provisional advisories for drinking water PFAS concentrations from 400 ng/L to 70 ng/L in 2016. Many states in the US are now contemplating limits ranging from 5–20 ng/L.

Over the past several decades, we have learned that exposures to a wide range of anthropogenic chemicals are associated with diverse deleterious health outcomes. There are critical windows of vulnerability to chemical exposure — such as the developing fetus during the third trimester of pregnancy, when the brain is developing most rapidly, and during the first several years of life when the body's immune programming is taking place. In their landmark 1996 book, *Our Stolen Future*, Theo Colburn, Dianne Dumanoski and John Peterson Myers brought together large amounts of scientific data linking declining human fertility with a rise in exposure levels to estrogen-like structures present in many common synthetic chemicals.[9] They put forward the so-called 'endocrine disruptor' hypothesis, arguing that hormone-like synthetic compounds were taking a heavy toll on humans and wildlife, interfering with the organism's natural chemical signaling pathways. We now also know that exposure to various forms of arsenic (As), PFAS and polychlorinated biphenyls (PCBs) is associated with impaired immune health, while PFAS and other synthetic organic compounds impair fat metabolism, potentially contributing to a growing global obesity epidemic. Linda Birnbaum, the recently retired Director of the US National Institute of Environmental Health Sciences, points out that the environment is suspected to be the primary cause for recent increases in the incidence of many chronic diseases in the US population, because shifts in lifestyle, diet and behavior patterns have not shown parallel trends.[10]

A major problem in regulating the production and release of toxicants into the environment has been the reactionary rather than proactive approaches to management that are still pervasive today. This began with the public outrage following Rachel Carson's description of the impacts of indiscriminate pesticide use in *Silent Spring* (1962),[11] which catalyzed the formation of the US EPA in 1970 under President Richard Nixon. Yet, despite its mission to 'protect human health and the environment,'[12] and an innovative mandate to study the health effects of new chemicals, progress by the EPA has been slow. In 1976, the first iteration of the Toxic Substances Control Act grandfathered 60,000 chemicals already in use, effectively considering them as safe and exempting them from further scrutiny. This law was finally revisited by the US Congress in 2016, with the Frank R. Lautenberg Chemical Safety for the 21st Century Act.[13] Similarly, the Clean Air Act (US) and the Clean Water Act, both seen as milestone achievements of environment legislation, have only focused on regulating a handful of specific toxicants. Only eighty-three contaminants are regulated in drinking water as part of the Safe Drinking Water Act and no new chemicals have been added since the law was promulgated in 1974. And sometimes, hard-won progress has been reversed. For example, long-standing US regulations on the emission of Hg from coal-fired utilities (the largest remaining source in the country) were rolled back by President Trump in 2019. If we are unable to regulate toxicants like Hg, well-established to pose public health risks with societal costs in the many billions of dollars, effective regulation for other chemicals seems nearly impossible.

On the international front, chemicals management has been similarly reactive in nature. Negotiations toward the establishment of the Stockholm Convention on Persistent Organic Pollutants (POPs) began in 2001, and the global treaty entered into force in 2004 with 128 signatory nations. However, only twelve POPs were initially included in the agreement (the 'dirty dozen'), and only nine were added after the first update in 2009. The global treaty on Hg (Minamata Convention on Mercury) entered into force in 2017, but progress has been slow in establishing international agreement on how emissions reductions will be accomplished. Although these agreements represent major accomplishments in international policy, their utility for reducing ubiquitous exposures to toxic substances has so far been limited.

To address this problem, the European Union put forward an innovative approach to chemical management in 2007 known as REACH: Registration, Evaluation, Authorization and Restriction of Chemicals.[14] This regulatory framework places the burden of proof for harm associated with chemical substances on manufacturers, and advocates for a more precautionary approach than used in North America. However, enforcement has proven to be challenging, and the general consensus among experts is that the ambitions of the regulations have not matched their accomplishments to date.

As we look to the future, advocacy on behalf of communities and public outrage remain the most effective and timely method for enacting changes in chemical use and releases. As a graduate student in the late 1970s, Arlene Blum, the famous mountaineer and founder of the Green Science Policy Institute, reported high levels of carcinogenic brominated flame retardants in children who wore treated pajamas, leading to the first regulations for these chemicals. More recently, public attention has been mobilized around the potential health impacts of PFAS, which have contaminated the drinking water of hundreds of communities across the US. Work by Rob Bilott, an attorney who litigated DuPont on behalf of the community in Parkersburg, West Virginia, which was affected by the company's pollution of that area, has made these odd-sounding chemicals a household concern across the country. His work was the subject of a major 2019 Hollywood film, *Dark Waters*, directed by Todd Haynes. Under increasing public pressure, states and the Federal US Government are now scrambling to respond with new regulations for drinking water, food contact materials and waste products that are used for biosolids. The main global manufacturer of PFAS in North America, 3M, voluntarily discontinued manufacturing the parent chemical to one of the most abundant PFAS found in the environment and humans between 2000–2002. This led to large and rapid declines in the concentrations of this chemical in the environment and human blood throughout North America and Europe, illustrating the benefits of coordinated decisive action on environmental releases. Unfortunately, however, PFAS have become the latest example of chemical whack-a-mole; one compound is phased out, only to be quickly replaced by another whose environmental properties and health consequences are largely unknown. The same game of chemical whack-a-mole has been played for different brominated flame retardants and plasticizers such as bisphenol-A in water bottles and other products. Each banned chemical is replaced by new compound that is initially assumed to be safe, but later discovered to be a regrettable substitution and problematic in its own right.

Addressing the global chemical experiment requires a new kind of thinking about environmental issues. First, current education and research still emphasize single disciplines. This isolates the chemical engineers responsible for creating new and better materials for society from the environmental toxicologist and health scientists who could screen for potentially deleterious effects prior to industrial use and widespread environmental release. Most chemical engineers are not currently trained in basic environmental science and risk assessment, yet this could be easily incorporated into standard teaching for undergraduates. Tools for screening potentially adverse impacts of new chemicals on human and ecological health have already been developed by various regulatory agencies, yet their full potential has yet to be realized. For example, the US EPA has developed a computational toxicology screening tool known as 'ToxCast' that uses high throughput screening assays to understand potentially adverse impacts of exposure for living organisms.[15] Use of this and other emerging screening tools would be simple and inexpensive prior to widespread use of chemicals in commerce.

The path toward sustainability in chemicals management is achievable. Tom Graedel from Yale's School of Forestry has shown that tracking the use of chemicals from manufacturing to disposal can improve conservation and optimize material flows.[16] Similarly, the movement towards a 'circular economy' has demonstrated to manufacturers that eliminating chemical releases through reuse rather than disposal can be profitable, while also providing good public relations. Positive notes for the future include the examples for Hg, Pb and PFAS, among others, where society has taken decisive action toward chemical management, and concentrations in humans and wildlife have dropped significantly shortly thereafter. Engineering innovations have produced emission control and waste treatment technologies that can virtually eliminate many of the chemicals of concern from our power plants and wastewater effluents. However, investment of societal resources in implementing such technologies remains a challenge that must be addressed to end our global chemical experiment.

Endnotes

1   J. N. Rauch and J. M. Pacyna, 'Earth's global Ag, Al, Cr, Cu, Fe, Ni, Pb, and Zn cycles', *Global Biogeochemical Cycles*, 2009, 23, GB2001, https://doi.org/10.1029/2008GB003376

2   'Better living through chemistry' is a variant of the famous DuPont advertising slogan, which was used by DuPont (a now defunct American conglomerate) in various forms from 1935 through to 1999.

3   See 'Land' by Navin Ramankutty and Hannah Wittman in this volume.

4   See 'Fresh Water' by Janet G. Hering in this volume.

5   See 'Air' by Jon Abbatt in this volume.

6   P. Grandjean and P. J. Landrigan, 'Developmental neurotoxicity of industrial chemicals', *The Lancet*, 2006, 368, 2167–78, https://doi.org/10.1016/s0140-6736(06)69665-7

7   R. L. Canfield, C. R. Henderson, D. A. Cory-Slechta, C. Cox, T. A. Jusko and B. P. Lanphear, 'Intellectual impairment in children with blood lead concentrations below 10 µg per deciliter', *New England Journal of Medicine*, 2003, 348, 1517–26, https://doi.org/10.1056/nejmoa022848

8   E. M. Sunderland, X. C. Hu, C. Dassuncao, A. K. Tokranov, C. C. Wagner and J. G. Allen, 'A review of the pathways of human exposure to poly- and perfluoroalkyl substances (PFASs) and present understanding of health effects', *Journal of Exposure Science & Environmental Epidemiology*, 2019, 29, 131–47, https://doi.org/10.1038/s41370-018-0094-1

9   T. Colborn, D. Dumanoski and J. P. Myers, *Our Stolen Future*. New York: Penguin Books, 1996.

10   L. S. Birnbaum, 'When environmental chemicals act like uncontrolled medicine', *Trends in Endocrinology and Metabolism*, 2013, 24, 321–23, https://doi.org/10.1016/j.tem.2012.12.005

11   R. Carson, *Silent Spring*. New York: Houghton Mifflin, 1962.

12   US Environmental Protection Agency, 'Our mission and what we do', https://www.epa.gov/aboutepa/our-mission-and-what-we-do

13   Available at https://www.congress.gov/114/plaws/publ182/PLAW-114publ182.pdf

14   European Chemicals Agency, 'Understanding REACH', https://echa.europa.eu/regulations/reach/understanding-reach

15   D. J. Dix, K. A. Houck, M. T. Martin, A. M. Richard, R. Woodrow Setzer and R. J. Kavlock, 'The ToxCast program for prioritizing toxicity testing of environmental chemicals', *Toxicological Sciences*, 2007, 95, 5–12, https://doi.org/10.1093/toxsci/kfl103

16   T. E. Graedel, 'On the concept of industrial ecology', *Annual Review of Energy and the Environment*, 1996, 21, 69–98, https://doi.org/10.1146/annurev.energy.21.1.69

Land

—

Navin Ramankutty and Hannah Wittman

Our planet is two-thirds ocean, yet we call it Earth; perhaps because we have created our homes on the relatively small fraction of the surface that is covered by land. The land is where we grow our crops, graze animals, obtain forest products for housing and medicinal needs, consume and distribute marine resources and earn our livelihoods. Land use is a primary manifestation of a socio-ecological metabolism that takes up, transforms and expresses energy in the service of human culture, technological progress and capital accumulation.[1] In the age of the Anthropocene, human-driven land use has a bigger impact on the global environment than almost any other activity.

Humans have been modifying the Earth's land surface since time immemorial as part of nomadic and then increasingly sedentary and complex agricultural societies. Even the soil — the foundation of humanity's survival — has a deep anthropogenic history; human practices of forest burning and fertilization of agricultural lands have long marked the human domination of Earth's land surface. Today, about 40% of the world's ice-free land surface is used for growing crops or grazing animals.[2] By comparison, urban areas make up less than 5% of Earth's surface, yet this small area is home to over half the human population.[3] Human influence is also evident in much of the remaining lands that have not been cleared, with estimates suggesting that more than three quarters of the world's lands bear the footprint of our species.[4]

   https://doi.org/10.11647/OBP.0193.22

To fully appreciate human impacts on Earth's surface, we can look to past drivers of global land use change, as we work to shift our practices towards a more sustainable future.[5] During the first stage of human history (the Paleolithic age, between about two million and ten thousand years ago), the use of stone tools and the control of fire enabled humans to migrate from their origins in East Africa to Eurasia, Australia and the Americas. The use of fire by Palaeolithic hunters changed landscapes and was also partly responsible for the extinction of megafauna. Observations of human-induced landscape burning can be traced to antiquity, as in a Carthaginian reference to western Africa around 500 years BC: '...during the day, we saw nothing but forests, but by night many burning fires ... we saw the land at night covered with flames. And in the midst there was one lofty fire, greater than the rest, which seemed to touch the stars'.[6]

The next stage of human history began with the domestication of plants and animals roughly 10,000 years ago.[7] This Neolithic Revolution occurred in several places around the world; first in Mesopotamia, China, eastern North America, New Guinea and the Sahel, and later on in Mesoamerica and the Andes. The advent of sedentary agriculture exerted further profound impacts on local landscapes. Both Plato and Aristotle commented on the soil erosion and deterioration of the hills and mountains of Greece. In describing the regional landscape, Plato wrote in *Critias*, 2400 years ago, that '...what now remains ... is like the skeleton of a sick man, all the fat and soft earth having wasted away, and only the bare framework of the land being left'.[8]

Fast forward a couple of thousand years, and we see that the most recent phase of human history has been marked by the human appropriation of energy stored in fossil fuels.[9] This began roughly 300 years ago, and was characterized by the rise of global trade networks, the advent of Industrial Revolution technologies and the dominance of capitalism. During this period, the extent and pace of human activities on the land surface accelerated drastically, as the global expansion of agriculture followed the development of human settlements and the world economy.[10] In 1700, large-scale agriculture was mainly confined to the Old World, in Europe, India, China and Africa. European colonization, a violent spatial 'fix' to the unmet food and energy needs of new industrial cities, created new settlement frontiers in the Americas, Australasia and South Africa (and was also responsible for cultural and linguistic genocide across several continents). At the same time, Russians moved east in the Former Soviet Union. The impact of this economic and agricultural development on Earth's surface has been dramatic. During the past three centuries, four times more land has been converted for human use than during all of prior human history.

During the past half-century, since the first Earth Day in 1970, agricultural expansion has taken a more complex turn. Agricultural frontiers in the tropics of Latin America, Southeast Asia and Africa have further expanded to meet the dietary needs of wealthy consumers in global cities, while other regions have seen abandonment of agriculture followed by regrowth of forests. Growing awareness of the importance of forest conservation for biodiversity and climate change mitigation has also led to an increase in protected areas across the globe; although rapid deforestation continues in many regions driven by global commodity markets for beef, soy, palm oil and sugar.[11] These shifts across the globe also correspond to the concentration of land ownership in higher-income countries, where aging farming populations struggle with farm succession and where government support for agriculture is concentrated in a few commodities such as maize, soy and wheat. At the same time, in lower-income countries, farms are becoming fragmented and are struggling with inequitable access to agricultural infrastructure.

Social and political trends over the past fifty years have driven a shift away from expansion of agricultural lands, toward more intensified use of existing areas. Although land clearing and deforestation continues in the tropics today, the total rate of cropland expansion over the last fifty years has been slower than in the previous two and half centuries. Yet, despite reduced clearing rates and reduced agricultural land area per-capita, global agricultural lands have continued to provide food and other agricultural products for the rapidly rising human population.

The apparent paradox of increased food production on a smaller per-capita land surface is a direct result of the so-called Green Revolution; the suite of technologies that enabled the yields of a small number of staple crops to increase rapidly since the 1950s.[12] The development of the Haber-Bosch process in the early twentieth century was particularly important, as it permitted the synthesis of nitrogen fertilizer from the plentiful (though biologically-inert) nitrogen in the atmosphere. This discovery was a major agricultural breakthrough, as nitrogen is a major limiting nutrient in soils. At the same time, scientific advances improved understanding of plant genetics and physiology, and their relationship to crop performance. Plant breeders were supported by both governments and private industry to develop new high yielding varieties of maize, wheat and rice that were able to effectively utilize the synthetic fertilizers that were rapidly deployed in the 1950s and 1960s, mainly in Latin America and Asia. These developments, coupled with low (subsidized) energy costs, allowed more capitalized farmers to efficiently exploit (and over-exploit) their soil and groundwater resources. Over the 1961–2014 period, the global area equipped for irrigation doubled, from 0.16 billion hectares (12% of cropland) to 0.33 million hectares (21% of cropland). As a result of new seed varieties, and additional nutrient, water and fossil energy inputs, per capita cereal production increased by 30% between 1961 and 2014.[13]

At the same time, public investment in just a small number of food crops has led to a rapid decline in agro-biodiversity. During the twentieth century, more than 90% of crop genetic diversity was lost to agriculture as farmers were encouraged by markets and public policies to shift their land use to fewer, higher-yielding varieties. The three plant species — rice, maize and wheat — that were the primary focus of the Green Revolution now contribute nearly 60% of the world's plant-based food supply. While urban consumers enjoy almost unlimited choice of foods in the supermarket (if they can afford it), just a few ingredients — wheat, maize and now soy — are found in most processed foods and nearly every meal of a typical North American diet.

The Green Revolution has been deemed a massive success by some scholars and policymakers, considering the overall increase in total global crop production.[14] Indeed, total calories produced increased more than 30% from 1961 to 2013, reaching an average of 2884 daily kcal per person, which is more than enough to meet the average minimum daily requirements of every person on Earth.[15] However, these calories are not distributed equally across the global food economy. The number of undernourished in the world remains unacceptably high at approximately 820 million people, and 2 billion suffer from micronutrient deficiencies. The global reliance on a few crops for energy is another primary reason for the human nutrition gap, with some 84% of global calories coming from just seventeen crops. On the flip side, excess and 'empty' calorie consumption has resulted in a global obesity epidemic, with around 37% of the world's population now overweight, carrying a heavy burden of non-communicable diseases such as diabetes, heart disease and certain cancers.

The unintended environmental consequences of the Green Revolution have burdened farmers, consumers and governments with new challenges in the management of sustainable land and food systems.[16] The clearing of forests and grasslands and the use of chemical fertilizers and pesticides have made agriculture the biggest driver of global biodiversity loss. Agriculture, forestry and other land uses also contributed nearly a quarter of global anthropogenic greenhouse gas emissions during the 2007–2016 period,[17] with 13% of carbon dioxide emissions resulting from tropical forest clearing, 44% of methane emissions from rice paddy cultivation, livestock enteric fermentation and manure, and 82% of all anthropogenic nitrous oxide emissions from excess fertilizer application (which quadrupled during 1961–2014). Moreover, irrigation now represents 70% of fresh water withdrawals around the world, and nitrogen and phosphorus loss from agricultural fields is the predominant driver of inland and coastal eutrophication.

The environmental impacts of agriculture and other associated land-use changes are also coupled to a suite of socio-economic factors. Poverty is a primary cause for the continued prevalence of malnutrition despite sufficient caloric availability at the national and global levels. A majority of the world's malnourished are poor farmers who are often hampered by the lack of land or secure land tenure, and an inability to acquire seeds or inputs to maintain soil fertility. Their poverty also creates a lack of resilience in the face of episodic losses resulting from extreme weather disasters (which may be increasing in frequency and intensity on a warming planet). Moreover, those farmers who are net producers of food (selling more than they buy) are adversely affected by low market prices in a globalized world. On the other hand, the urban poor, who are net buyers of food, are severely affected by food price hikes. The ability of individuals to access food is thus affected by the balance between income and food price, as well as their position in the global food system.

All in all, more than two billion people across the world remain malnourished, even as the expansion of global agriculture is arguably the single most important driver of global environmental degradation. One of the key United Nations Sustainable Development Goals is to 'End hunger, achieve food security and improved nutrition and promote sustainable agriculture'.[18] How do we reduce food insecurity, feed the additional two to three billion people of the future, lower the environmental footprint of agriculture and make it more resilient to climate change — all at the same time?

This is, no doubt, a daunting challenge, but a number of wide-ranging solutions are now on the table. Many scholars recognize the need for continued increases in crop yields, particularly in areas characterized by high poverty rates with limited agricultural infrastructure such as irrigation, roads and markets. They call for sustainable intensification to produce more food at lower environmental costs, and argue that new agri-food technologies, such as precision agriculture and genetically modified (GM) foods are important components of this pathway. But others challenge this 'productivist paradigm', pointing out that increased food production is the wrong objective given the existing market failures that result in the poor global distribution of calories, and the fact that lower yields often reflect a lack of resources rather than technology. These scholars argue for a focus on food sovereignty, which advocates for growers and eaters to work together, along with scientists and the public sector, to develop regionally-adapted solutions for more equitable and ecological farming systems.[19] These systems employ agro-ecological methods like crop diversification and integration with animal agriculture to address soil nutrient deficiencies, while also contributing to dietary diversity and improved food security.

Looking ahead, numerous questions arise. There are many voices with differing opinions on how the world's land should be used and managed. Some believe in a top-down, regulatory approach, while others rely on the power of markets to determine what should be best done with the land. Still others suggest that local communities have the greatest incentives to protect their land-based resources, and should have the autonomy to make those decisions locally, recognizing that some voices have more influence than others. Further, we must understand that the current landscape results from the often unequal and unjust histories of past land ownership and use. Who decides how the land and food system challenges are framed and negotiated — from a local to a global scale — is as important as what happens on the land. There is no escaping the fact that global environmental challenges cannot be addressed without considering land-based solutions. Any such solutions will inherently involve trade-offs and inequities, which must be carefully considered in the design and implementation of effective, efficient and equitable policies for the future.

Endnotes

1. K. H. Erb, 'How a socio-ecological metabolism approach can help to advance our understanding of changes in land-use intensity', *Ecological Economics*, 2012, 76, 8–14, https://doi.org/10.1016/j.ecolecon.2012.02.005

2. J. A. Foley et al., 'Solutions for a cultivated planet', *Nature*, 2011, 478, 337–42, https://doi.org/10.1038/nature10452

3. United Nations Department of Economic and Social Affairs, *Revision of World Urbanization Prospects*, 2018, https://www.un.org/development/desa/publications/2018-revision-of-world-urbanization-prospects.html

4. O. Venter et al., 'Sixteen years of change in the global terrestrial human footprint and implications for biodiversity conservation', *Nature Communications*, 2016, 7, 12558, https://doi.org/10.1038/ncomms12558

5. C. Redman, *Human Impact on Ancient Environments*, Tucson: University of Arizona Press, 1999, 288.

6. Hanno, *The Periplus of Hanno: A Voyage of Discovery down the West African Coast, by a Carthaginian Admiral of the Fifth Century B.C.*, trans. by W. H. Schoff, Philadelphia: The Commercial Museum, 1912, 4–5.

7. B. L. Turner II and S. McCandless, 'How humankind came to rival nature: A brief history of the human-environment condition and the lessons learned, in earth system analysis for sustainability', in *Dahlem Workshop Report No. 91*, ed. W. C. Clark et al., Cambridge, MA: MIT Press, 2004, 227–43.

8. Plato, *Plato in Twelve Volumes*, Vol. 9, trans. by W. R. M. Lamb, Cambridge, MA: Harvard University Press, 1925, 111.

9. Turner II and McCandless, 'How humankind came to rival nature', 2004.

10. N. Ramankutty, Z. Mehrabi, K. Waha, L. Jarvis, C. Kremen, M. Herrero and L. H. Rieseberg, 'Trends in global agricultural land use: Implications for environmental health and food security', *Annual Review of Plant Biology*, 2018, 69, 789–815, https://doi.org/10.1146/annurev-arplant-042817-040256

11. See 'Forests' by Sally N. Aitken in this volume.

12. P. L. Pingali, 'Green Revolution: Impacts, limits, and the path ahead', *Proceedings of the National Academy of Sciences of the United States of America*, 2012, 109, 12302–08, https://doi.org/10.1073/pnas.0912953109

13. Ramankutty et al., 'Trends in global agricultural land use', 2018.

14. Pingali, 'Green Revolution', 2012.

15. Ramankutty et al., 'Trends in global agricultural land use', 2018.

16. D. Tilman, 'The greening of the green revolution', *Nature*, 1998, 396, 211–12, https://doi.org/10.1038/24254

17. IPCC, 'Technical Summary', in *Climate Change and Land: An IPCC Special Report on Climate change, Desertification, Land Degradation, Sustainable Land Management, Food Security, and Greenhouse Gas Fluxes in Terrestrial Ecosystems*, ed. H.-O. Pörtner et al., Geneva: IPCC, 2019, 37–74, at 41, https://www.ipcc.ch/site/assets/uploads/sites/4/2019/11/03_Technical-Summary-TS.pdf

18. The Sustainable Development Goals Report documenting these goals is available at https://unstats.un.org/sdgs/report/2016/

19. H. Wittman, 'Food sovereignty: A new rights framework for food and nature?', *Environment and Society: Advances in Research*, 2011, 2, 87–105, https://doi.org/10.3167/ares.2011.020106

Oceans 2020

—

David M. Karl

I was born in 1950, at the beginning of the second half of the twentieth century, and just five years after the end of the Second World War. It was a time when rapid advances in science, technology and medicine were fundamentally changing the human experience on Earth, and not all for the better. As an eighteen-year-old high school senior, I read Paul Ehrlich's *The Population Bomb*,[1] a dire warning to humanity about the state of the global environment, with apocalyptic predictions about the future. I knew right then that we needed all hands on deck to deal with an impending crisis of our own making.

Two years later, in 1970, the first Earth Day was created to honor our planet and its diverse ecosystems. This event had its roots on college campuses across the United States under the leadership of then US Senator Gaylord Nelson, and it gained momentum and energy from the ongoing Vietnam War protests. I, too, was engaged in anti-war demonstrations, while at the same time pondering my own future. I was drawn to studies of environmental science and ecology, with a keen interest in aquatic habitats. Growing up in Buffalo, New York, on the heavily polluted shores of Lake Erie, I saw, firsthand, how industrial and municipal wastes were threatening the survival of all forms of aquatic life. In the late 1960s, oily slicks of toxic chemicals repeatedly caught fire in tributaries of the Lake, including the Buffalo and Cuyahoga Rivers.[2]

By 1971, environmental issues had already become more mainstream, and on April 21 of that year, I was invited to participate in 'Survival Day' — our neighborhood equivalent of

 https://doi.org/10.11647/OBP.0193.23

the second Earth Day. My brother, Tom, was a high school teacher, and he helped organize this first-ever, local environmental event. Speakers included distinguished representatives from advocacy organizations, Health Department officials, university professors, as well as industry representatives from Allied Chemical Company, Bethlehem Steel and Niagara Mohawk — the regional energy provider. As a college student, without any fancy title or affiliation, I was listed in the program as 'Concerned resident of planet Earth'. And I was. My talk on the *Biogeochemical Effects of Bethlehem Steel on Lake Erie* was my first formal foray into the environmental movement — and from there, I never looked back. University degree in hand, I left Buffalo to begin my new career as an oceanographer, with an interest in understanding microbial processes in the deep blue sea, far removed from the influences of humankind. Or so I thought.

E xploration of the global ocean dates back many centuries, well before the European-led age of discovery or the Polynesian seafarers of the Pacific Ocean. Between the fifteenth and seventeenth centuries, explorers ventured far from their coasts to discover new lands, resources and wealth. The doctrine at that time, *mare clausum*, was one of exclusive ownership and navigational rights, even on the high seas. This policy changed radically in 1609 with the publication of Hugo Grotius' new principle of *mare liberum*, freedom of the sea,[3] which redefined the rules of sovereignty, and facilitated the growth of maritime activities including colonial expansion, commerce and scientific research. The great worldwide voyages of James Cook, Charles Darwin and James Clark Ross were partially motivated by scientific inquiry, but most scholars would agree that the four-year (1872–1876) worldwide voyage of HMS *Challenger* marked the true beginning of modern oceanography.

The first half of the twentieth century witnessed the expansion of marine laboratories and oceanographic research vessels worldwide. In 1927, the president of the US National Academy of Sciences (NAS) appointed a committee on oceanography to consider the worldwide scope of the discipline. The work of this committee was published over the next decade by Henry Bryant Bigelow and Thomas Wayland Vaughan,[4] among others, and served as the background and motivation for the classic treatise on oceanography by H. U. Sverdrup, M. W. Johnson and R. H. Fleming.[5] Key legacies of the NAS committee on oceanography were the establishment, in 1931, of the Woods Hole Oceanographic Institution, the creation of the Office of Naval Research, in 1946, and the establishment of public funding, in 1950, from the National Science Foundation for oceanographic research.

By the early 1950s, planning was underway for what would eventually become the 1957–1958 International Geophysical Year (IGY), a comprehensive study of Earth and its oceans involving 30,000 scientists from sixty-six countries. One of the most important achievements of the IGY was the establishment, by the oceanographer Roger Revelle and his colleague Charles David Keeling, of a laboratory atop the Hawaiian volcano, Mauna Loa, for continuous measurements of atmospheric carbon dioxide (CO_2). These measurements soon revealed a regular seasonal pattern of CO_2, which reflected the net balance between planetary photosynthesis and respiration along with exchanges of atmospheric CO_2 with the upper ocean. After a few years, Keeling was able to document a small, but systematic, rise in the average atmospheric CO_2 from year to year, resulting primarily from fossil fuel combustion. Other CO_2 sampling sites were soon established at strategic locations worldwide, along with the creation of an international network of ocean weather ships, which, at its peak, included twenty-two Atlantic and twenty-four Pacific Ocean stations collecting oceanographic and meteorological observations. In the decades that followed, these long-term data sets would prove to be critical for detecting anomalous ocean conditions and for establishing baselines against which future ocean states could be compared.

In the decade that followed Keeling's early CO_2 measurements, scientific progress towards understanding the global oceans began to accelerate. In the US, the Stratton Commission, appointed by President Lyndon B. Johnson, developed a national ocean action plan[6] based on their comprehensive, long-range assessment of marine health and necessary research activities. Although the country was preoccupied with the Vietnam War and the developing space program, many of the recommendations of the Stratton Commission were eventually enacted, including the creation of the National Oceanic and Atmospheric Administration (NOAA) in 1970. Other important outcomes included the creation, in 1972, of the Coastal Zone Management Act, National Marine Sanctuaries Act, Marine Mammal Protection Act, and, in 1976, the Magnuson–Stevens Fishery Conservation and Management Act.

Throughout the 1970s, funding for oceanographic research increased significantly, with the International Decade of Ocean Exploration (IDOE) and other programs that stimulated large-scale, multi-disciplinary oceanographic research. These international programs began to view the ocean as an integrated system, and included studies to preserve the marine environment, improve environmental forecasting, develop advanced ocean monitoring systems and facilitate the worldwide exchange of oceanographic data.

Progress accelerated through the 1980s, with the establishment of the International Geosphere-Biosphere Programme (IGBP). In 1988, under the auspices of the Joint Global Ocean Flux Study (JGOFS) program, two ocean time-series stations were established: one in the North Atlantic near Bermuda, and the other in the North Pacific near Hawaii. Since that time, near-monthly observations have been conducted at these two open ocean sites, building on other existing time-series stations, including the location of the former weather ship at Ocean Station Papa in the Gulf of Alaska, where oceanographic measurements have been made since the 1960s.

The early ship-based oceanographic surveys and time-series stations were critical for providing important baseline observations. But given the vastness of the planet's oceans, these measurements could not even hope to cover all of Earth's marine waters. Fortunately, just as ship-based oceanographic programs were ramping up, ocean science entered into the satellite age. In 1978, the coastal zone color scanner (CZCS) was launched on the Nimbus 7 satellite, providing the first dedicated imagery of ocean color, which was used to measure the concentration of photosynthetic plankton in marine surface waters. Initially designed as a one-year proof of concept, the CZCS mission ran until 1986, and yielded unprecedented information on the spatial and temporal patterns of biological productivity across the oceans. Since that time, improved satellite remote-sensing of ocean color, temperature, salinity, wind, sea level, sea ice and other key environmental variables has revolutionized our understanding of oceanographic processes on regional-to-global scales.

The development of marine science over the past half-century has coincided with a period of unprecedented human impacts on our oceans. Back in 1970, I (and others) believed that the oceans' vastness would serve to buffer any potential anthropogenic perturbations. Today we know that this is not the case. On a global scale, the oceans have warmed appreciably, as documented by successive reports of the Intergovernmental Panel on Climate Change (IPCC). This warming is largest near the sea surface (between 0 and 75 m depth), with a temperature increase of 0.11°C per decade since the first Earth Day, fifty years ago. This observed temperature trend has been documented with *high confidence*, in IPCC parlance.[7]

Rising ocean temperatures have both direct and indirect effects on marine ecosystems. For some species living at or near a temperature optimum, rising temperatures may approach or exceed physiological thermal tolerances, resulting in mass species migration. Indeed, the distributions and abundances of many marine organisms have already shifted poleward, or to deeper and colder waters as a result of ocean warming. Based on a fifty-year record from the North Atlantic Ocean, the range limit of warm water copepods (small planktonic animals eaten by fishes) has shifted north by ten degrees latitude. This poleward shift had led to seasonal mismatches in the growth cycles of primary producers, zooplankton grazers and predatory fishes, with significant ecosystem effects. Warming of polar regions is especially concerning owing to the narrow temperature ranges of many species, and the lack of colder water habitat refuges. In addition, organisms with limited ability to migrate, including tropical corals, face habitat loss, thermal-induced bleaching (loss of photosynthetic symbionts) and, possibly, extinction. The rate of temperature change in many marine ecosystems is unprecedented, so genetic adaptation and evolution are often unable to keep up.

The warming of surface ocean waters also has significant indirect effects on marine ecosystems. Rising temperature increases the density difference between the upper sunlit layers and the deeper, nutrient-rich waters below. This, in turn, reduces vertical mixing of water masses and the supply of nutrients for photosynthesis. This 'stratification' explains the deep blue color of tropical oceans, where low productivity ocean 'deserts' result from limited nutrient supply into the warm surface waters. Satellite observations of ocean color over the past several decades have revealed a significant global expansion of these oceanic deserts, and this trend is expected to continue in a warming future.

Ocean warming is also leading to the loss of sea ice as a critical habitat in high latitudes,[8] and changes in critical pathways of ocean circulation. Most notably, warming surface waters in the subpolar North Atlantic (near Greenland) could act to slow down the formation of cold and salty water masses, which sink into the ocean interior carrying nutrients and dissolved gases throughout the ocean depths. These sinking waters are closely coupled to the northward flow of the warm Gulf Stream current, which transports large amounts of heat to northern Europe. Should this circulation slow or even stop (as it appears to have done in the geological past), Europe could, counter-intuitively, experience less warming, or even some cooling, into the future. Moreover, sluggish ocean circulation, combined with lower gas solubility in warm waters is acting to decrease the concentration of dissolved oxygen (O_2) over much of the ocean interior. This is particularly problematic in regions that are naturally low in O_2, including large parts of the North Pacific, and in regions where additional anthropogenic nutrient inputs fuel excessive microbial O_2 consumption. In extreme cases, low O_2 conditions in some coastal sites have created so-called 'dead zones' leading to massive mortality of fishes and bottom-dwelling invertebrates.

Beyond its effect on global temperature, increasing atmospheric CO_2 concentrations are altering the chemistry of the surface ocean in a way that is negatively impacting many marine organisms. To date, approximately 25% of the CO_2 that has been emitted by human activities has been absorbed by the surface ocean.[9] On the face of it, this CO_2 enrichment might be expected to benefit ocean life by stimulating marine photosynthesis. The reality, however, is more complex. Perhaps most importantly, there is the problem of ocean acidification, which is a direct result of increasing ocean CO_2 levels, since dissolved CO_2 reacts with seawater to produce carbonic acid. Between 1988 and 2018, surface ocean acidity at the Hawaii Ocean Time-series Station ALOHA increased by 14%. This might not sound like a large change, but even very small changes in acidity can have profound consequences for the growth of many organisms. Marine species that produce calcium carbonate as a support structure, shell or skeleton (for example, shellfish and reef-building coral) will need to invest additional metabolic energy to form calcium carbonate. Furthermore, exposure of calcified structures to more acidic seawater will weaken or even completely dissolve the life-supporting calcium carbonate structures. In addition, water-breathing fishes may be impacted by increasing acidity in their bloodstream, creating an additional physiological stress beyond ocean warming and deoxygenation.

As we reflect on the past fifty years of ocean change, we must also look to the future. The United Nations has proclaimed a Decade of Ocean Science for Sustainable Development (2021–2030) to address current and emerging threats to global marine ecosystems, including the insidious pollution by anthropogenic micro-plastics.[10] My generation is solely responsible for the global growth of plastics, a successful subsidiary of the oil industry. Of the long list of insults to marine ecosystems, plastic pollution might be the 'low-hanging fruit' for successful remediation, simply by enacting bans on single-use plastics, and increasing the effectiveness of recycling programs. The 2019 G20 summit in Tokyo released a joint declaration on the critical need for marine conservation, and the elimination of plastic waste. Equally concerning is the accelerated pace of proposals to mine deep ocean metal deposits. Since its inception in 1982, the International Seabed Authority has issued numerous leases for mineral exploration in the deep sea. The first commercial operation off Papua New Guinea had planned to mine mineral-rich hydrothermal vents from depths of 1.5–2 km, but financial problems forced it to file for bankruptcy in 2019. However, other nations and companies continue to map resources within their leased regions of the seabed, despite vocal and well-informed opposition by oceanographers and marine conservationists. The potential impacts on deep sea habitats are well documented, but these are pitted against the growing need for raw materials to sustain our current standard of living and future population growth. Who will referee the competing interests of humankind versus nature? And who will win?

In the face of significant challenges, we can take solace, and perhaps even inspiration, from the diverse marine microbial assemblages that have thrived on our planet for billions of years. These microorganisms possess enormous genomic potential and metabolic flexibility, and this has provided them with resilience in the face of environmental change. In the end, marine microbes will survive and adapt to climate change, although it is unclear how humankind will fare. Despite an ever-growing knowledge base concerning the sea around us built on observations, measurements and computer models, the ocean is still grossly under-sampled. Consequently, major uncertainties still exist regarding climate change impacts on the ocean and its inhabitants. Human influence on climate is indisputable and accelerating, and now, more than ever, we need to embrace a holistic view of the coupled Earth systems. Basic science is critical, but so too is fact-based education, aggressive advocacy for our planet and effective action.

Endnotes

1. P. R. Ehrlich, *The Population Bomb*, New York: Ballantine Books, 1968.

2. See also 'Fresh Water' by Janet G. Hering in this volume.

3. H. Grotius, *Mare Liberum, Sive De jure quod Batavis competit ad Indicana commercia Dissertatio* [The Freedom of the Seas, or a Disputation Concerning the Right Which Belongs to the Dutch to take part in the East Indian Trade], Leiden: Lodewijk Elzevir, 1609.

4. H. B. Bigelow, *Oceanography: Its Scope, Problems and Economic Importance*, New York: Houghton Mifflin Company, 1931; T. W Vaughan et al., *International Aspects of Oceanographic Data and Provisions for Oceanographic Research*, Washington, DC: National Academy of Sciences, 1937, https://doi.org/10.5962/bhl.title.16994

5. H. U. Sverdrup, M. W. Johnson and R. H. Fleming, *The Oceans: Their Physics, Chemistry and General Biology*, New York: Prentice-Hall, Inc., 1942.

6. J. Stratton et al., *Our Nation and the Sea: A Plan for National Action*, Washington, DC: Commission on Marine Science, Engineering and Resources. US Government Printing Office, 1969.

7. IPCC, 'Climate change 2013: The physical science basis', in *Contribution of Working Group 1 to the Fifth Assessment Report of the Intergovernmental Panel on Climate Change*, ed. T. F. Stocker, D. Qin, G.-K. Plattner, M. Tignor, S. K. Allen, J. Boschung, A. Nauels, Y. Xia, V. Bex and P. M. Midgley, Cambridge, UK: Cambridge University Press, 2013, 383–464, https://www.ipcc.ch/site/assets/uploads/2018/02/WG1AR5_all_final.pdf; IPCC, 'Climate change 2014: Impacts, adaptation and vulnerability, part A: Global and sectoral aspects', in *Contribution of the Working Group 2 to the Fifth Assessment Report of the Intergovernmental Panel on Climate Change*, ed. by C. B. Field et al., Cambridge, UK: Cambridge University Press, 2014, 1–1131, https://www.ipcc.ch/site/assets/uploads/2018/02/WGIIAR5-PartA_FINAL.pdf

8. See also 'Ice' by Julian Dowdeswell in this volume.

9. On oceanic uptake of CO_2, see 'Carbon' by David Archer in this volume.

10. See also 'Earth and Plastic' by Roland Geyer in this volume.

Earth and Plastic

—

Roland Geyer

A few months after the first Earth Day, in summer 1970, a handful of researchers at the Massachusetts Institute of Technology (MIT), began to work on a computer simulation that would forever change the way we think about the world. The research was ground-breaking in multiple ways. It used a novel mathematical modeling technique called system dynamics, and an equally novel approach, computer simulation, to study the interactions between human society and the natural environment on a global scale. In particular, the project examined what could happen when exponential growth in human population and economic output is confronted with the finite resources of planet Earth. Today, PhD students run much more complicated simulations on their laptops, but back then, this approach was in its infancy, and non-linear behavior of systems was neither well understood nor studied much.

Results from the MIT study were published in a 1972 book called *The Limits to Growth*, which contained dire prognoses about the consequences of continued exponential growth of the human population and the global economy.[1] The book sparked instant controversy and received fierce criticism, especially from economists. This was perhaps to be expected; claiming that business as usual would lead to global 'overshoot and collapse' is unlikely to make you popular. One widespread criticism was based on the erroneous interpretation of the study as predicting the depletion of non-renewable resources within a few decades — a

  https://doi.org/10.11647/OBP.0193.24

depletion that did not materialize. However, rather than generating any singular predictions, the computer simulations were conceived as a means of exploring many plausible if-then scenarios. Critics also commonly overlooked the fact that many model runs predicted overshoot and decline even when resources were assumed to be limitless. In these model runs, it was Earth's capacity to assimilate human wastes and emissions (represented as pollution in the model), rather than the supply of raw materials and fuel, that became the critical environmental constraint on the continued growth of the economy.

Today, nearly half a century after the MIT study, there is growing consensus that environmental pollution caused by wastes and emissions is of far greater concern than depletion of non-renewable resources. Fossil fuels are the perfect example for this general insight. Since the 1970s, there have been various predictions for the year in which global oil production would peak and then steadily decline. Early predictions were off, since they did not account for unconventional sources of oil, such as tar sands and shale oil, and novel technologies, such as horizontal drilling and hydraulic fracturing (fracking). While the controversy over 'peak oil' continues, it is largely irrelevant; the true environmental constraint on fossil fuel consumption is the amount of carbon dioxide (CO_2) released into the atmosphere, and the effects of this long-lived greenhouse gas on global climate.[2]

In 1970, global atmospheric CO_2 abundance was approximately 320 parts per million (ppm). In 2018 it reached 410 ppm, enough to cause potentially catastrophic climate change.[3] During the same period, global annual CO_2 emissions from fossil fuel combustion had increased from 14 to 34 billion metric tons, or Gigatons (Gt). According to British Petroleum's Statistical World Energy Review, global proved fossil fuel reserves at the end of 2018 were 1,730 billion barrels of oil, 197 trillion cubic meters of natural gas, and 1,055 Gt of coal.[4] 'Proved reserve' here means that the resource could be recovered with reasonable certainty under existing economic and operating conditions. The combined proved fossil fuel reserves contain well over 1,000 Gt of carbon. If all of this carbon were combusted and released into the atmosphere as CO_2, it would further raise the atmospheric CO_2 levels to a point that would make any efforts to avoid catastrophic climate change completely futile. In other words, we will have wrecked the climate long before we run out of fossil fuels.

Not all extracted fossil fuels are burned, though. Today, 14% of oil production and 8% of natural gas extraction is used to make petrochemicals, such as plastics, fertilizers and a multitude of other chemicals. Petrochemicals production as a whole experienced enormous growth since the end of the Second World War, and the rise of plastics, in particular, is its most visible manifestation. As a mass-produced material, plastics are barely seventy years old.[5] In 1970, the year of the first Earth Day, global annual production of plastic polymer resins, fibers and additives was 37 million metric tons, or megatons (Mt). In 2017, global annual plastics production had reached an astounding 438 Mt, an eleven-fold increase in less than fifty years. By the end of 2017, humankind had produced a total of 9.2 Gt of plastic. That is the equivalent mass of 900,000 Eiffel Towers, or 88 million blue whales, or 1.2 billion elephants. If spread out ankle deep as low density plastic waste, it would cover an area the size of Argentina, the eighth largest country in the world. The growth of global annual plastic production has been so large and sustained that half of all plastic ever made by humankind was produced in just the last thirteen years. In other words, in just a little more than the past decade alone, we have doubled the total amount of plastic ever made.

While some might regard the global rise of plastic as a fantastic economic success story, others see an environmental tragedy. Many plastic products are short lived — plastic toys, household items, or fast fashion made from synthetics, for example. But it is packaging that has the shortest lifetime of all plastic products. Packaging accounts for around 36% of plastic production, most of which is used once and then disposed of. As a result, much plastic becomes waste soon after it was produced, and plastic waste generation can thus be expected to closely track plastic production. Unfortunately, solid waste generation data are much harder to come by than material production data — clear evidence that we consider the generation of solid waste an inconvenience, and treat it as an afterthought.

We love buying new things, in alluring and convenient packaging, but we also seem to expect that the old stuff will just disappear once we throw it into our garbage bins. This may have been true at some point in the past, when the majority of our trash would rot or corrode away. It is certainly not true for the plastics we have made so far, since they do not biodegrade on any reasonable timescale. In fact, all of the plastic we have made and did not burn, or otherwise destruct thermally, is still present on this planet. This is estimated to be 86% of the plastic waste humankind has thus far generated. An estimated 6 Gt of plastic waste is therefore present somewhere on this planet: in landfills, or open dumps, or in the natural environment. Another estimated 3 Gt is currently in use and will become waste as soon as we're done with it, which won't be long.

While we can estimate how much plastic currently resides on the planet, we do not know where exactly it is. Conventional plastic polymers don't biodegrade, but become brittle and disintegrate into smaller and smaller pieces, which then disperse in the environment as so-called micro-plastics. Wherever we look for plastic we find it. Plastic has been found in ocean creatures of all sizes and trophic levels, from plankton and seabirds to fish and whales. It's on the ocean surface, in the water column, and on the world's beaches, river beds and ocean floor, including its deepest point, the Mariana Trench, more than 11 km below sea level. Plastic has been found in arctic sea ice, in snow, rain, tap water, bottled water and beer. In the year 2010 alone, 5 to 13 Mt of plastic entered the world's oceans from land due to littering or mismanagement of plastic waste.[6] Terrestrial plastic pollution has so far received less attention than plastic marine debris, but we know that plastic is also everywhere in the soil. In fact, due to its ubiquity, plastic has recently been proposed as a geological indicator of the proposed Anthropocene, the period in which many geological surface processes started to be dominated by humans. A sediment core taken off the coast of Southern California shows the first appearance of micro-plastic in its sedimentary depth layers around 1950, with a subsequent doubling about every fifteen years thereafter. The long-term consequences of such pervasive and near-permanent plastic pollution are unclear at this point, but there are many reasons to expect significant adverse ecological and human health effects.

Traditional approaches to solid waste management are unable to cope with the ever-growing amount of plastic waste. Developed economies rely on a mix of landfill, incineration and collection for recycling. Landfill of plastic is essentially permanent storage of the waste material. Apart from the land, money and overall effort required, landfill also raises concerns about the generation and emission of hazardous substances. On average, about 8% of finished plastic consists of so-called additives; complex chemicals such as plasticizers, flame retardants and stabilizers, some of which are known to be hazardous. Many developed economies are aiming to reduce landfill rates by increasing incineration and collection for recycling. While plastic incineration rates are high in many European countries, waste incineration remains unpopular in the United States. The environmental and health impacts of waste incinerators very much depend on their emission control technology, as well as incinerator design and operation. Recycling delays rather than avoids final disposal, unless it reduces virgin plastic production (i.e. synthesis of plastics from hydrocarbon building blocks). European countries and the US used to send up to 60% of their plastic waste collected for recycling to China and Southeast Asia, but these countries are accepting less and less of it. In addition, plastic recycling suffers from poor environmental and quality control or poor economics, since it has to compete against cheap and abundant supply of virgin plastic. It is estimated that only 9% of all plastic ever made has been recycled. Many developing economies lack solid waste management infrastructure and have high rates of mismanaged plastic waste.[7] Considering all of these challenges, it is perhaps not surprising that a sizeable fraction of our plastic ends up in the natural environment.

The virgin plastics industry frequently states that using plastic is actually environmentally beneficial, since it replaces heavier and more impactful materials, including metal and glass. This argument not only implies that the environmental impacts of plastic production, use and disposal are lower than those of alternative materials, but also that plastic is being used instead of these other materials. Unfortunately, global production of all human-synthesized materials has been increasing, so we're actually using plastic in addition to everything else, not instead of it. In the fifty years since the first Earth Day, global annual production of hydraulic cement increased seven-fold, while primary aluminum and crude steel production grew by factors of five and three, respectively. Along with our increasing use of various materials, solid waste generation in general is also increasing year over year. Producing and using more materials each year does not just mean that there will be more waste when these materials reach the end of their useful lives. Materials cause environmental impacts throughout their life cycles, such as ecosystem disturbance during extraction, and wastes and emissions all along their supply chains. What we throw into our garbage bins is just the tip of an ever-growing 'wasteberg'.

Material recycling, which has recently been repackaged as part of the 'circular economy', is unlikely to be a panacea.[8] Collection and reprocessing of solid waste into secondary material has its own environmental impacts. These impacts are typically much lower than those of making the same material from primary resources, like ores, but they are still significant. Recycling therefore only decreases environmental damage if it significantly reduces primary material production. So far this has not happened. The currently empty promise of recycling is perfectly illustrated by a petrochemical engineer, who stated that 'we passionately believe in recycling' while overseeing the construction of a brand-new, 'as big as you get' virgin plastic plant that will be fueled by abundant Marcellus Shale gas.[9]

Another proposed solution that is not going to work on a global scale is the so-called 'bio-economy', in which fuels and materials are made from biomass rather than non-renewable resources. Utilizing waste from agriculture and forestry for fuel and material production is certainly attractive, but there is nowhere near enough biowaste for a large-scale replacement of non-renewable fuels and materials. Instead, this would require vast amounts of dedicated crop production and thus agricultural land. It is now clear that the climate change and land use impacts of global food production are imposing significant stress on Earth's terrestrial ecosystems.[10] Imagine how much those impacts would increase if we produce biomass not just for our food and feed, but also to supply all our fuels and materials. To make things worse, some bioplastics and biofuels, such as polyhydroxyalkanoates (PHAs) and corn-based ethanol, have been shown to have greenhouse gas emissions that are similar or even higher than those of their fossil-based competitors. A global bio-economy would also massively increase other environmental impacts, such as eutrophication from fertilizer runoff, where excess of nutrients leads to harmful algal blooms and oxygen depleted 'dead zones' in water bodies.

At this point, the only meaningful path forward will have to include substantial reductions in the amount of materials we produce and use, unless we are willing to see further increases of CO_2 in the atmosphere, plastic in the oceans, nitrogen in our estuaries and coastal waters and so on. It is telling when the CEO of Recology, a major resource recovery company, publishes a newspaper op-ed entitled 'It is time to cut the use of plastics'.[11] I have no doubt that a large reduction in our material footprint is compatible with a good life. If anything, maintaining the latter will require the former, since the relentless growth of the global economy seems to be finally hitting the environmental limits of this big, but finite, planet. This brings me back to *The Limits to Growth* study, conducted almost fifty years ago. One final error of its critics is the belief that history proved it wrong a long time ago. But the standard scenario in the report made projections well beyond the year 2050, finding that global population would peak around that time and decline sharply thereafter. It remains to be seen whether history will falsify this grim prediction.

Endnotes

1. D. Meadows, J. Randers and W. Behrens, *The Limits to Growth*, New York: Universe Books, 1972.

2. For an outline of CO_2's history, see also 'Carbon' by David Archer in this volume.

3. Global Monitoring Division of the National Oceanic & Atmospheric Administration (NOAA), US Department of Commerce, 'Trends in atmospheric carbon dioxide', https://www.esrl.noaa.gov/gmd/ccgg/trends/

4. British Petroleum (BP), *BP Statistical Review of World Energy 2019*, London: BP, 2019, https://www.bp.com/content/dam/bp/business-sites/en/global/corporate/pdfs/energy-economics/statistical-review/bp-stats-review-2019-full-report.pdf

5. R. Geyer, J. Jambeck and K. Lavender Law, 'Production, use and fate of all plastics ever made', *Science Advances*, 2015, 3, e1700782, https://doi.org/10.1126/sciadv.1700782

6. J. Jambeck, R. Geyer, C. Wilcox, T. Siegler, M. Perryman, A. Andrady, R. Narayan and K. Lavender Law, 'Plastic waste inputs from land into the ocean', *Science*, 2015, 347, 768–71, https://doi.org/10.1126/science.1260352

7. D. Hoornweg and P. Bhada-Tata, 'What a waste: A global review of solid waste management', *Urban Development Series Knowledge Papers*, 15, Washington DC: World Bank, 2012, http://

documents.worldbank.org/curated/en/302341468126264791/What-a-waste-a-global-review-of-solid-waste-management

8. R. Geyer, B. Kuczenski, T. Zink and A. Henderson, 'Common misconceptions about recycling', *Journal of Industrial Ecology*, 2016, 20, 1010–17, https://doi.org/10.1111/jiec.12355

9. M. Corkery, 'A giant factory rises to make a product filling up the world: Plastic', *New York Times*, 12 August 2019, https://www.nytimes.com/2019/08/12/business/energy-environment/plastics-shell-pennsylvania-plant.html

10. IPCC, *Climate Change and Land: An IPCC Special Report on Climate change, Desertification, Land Degradation, Sustainable Land Management, Food Security and Greenhouse Gas Fluxes in Terrestrial Ecosystems*, ed. H.-O. Pörtner et al., Geneva: IPCC, 2019, https://www.ipcc.ch/report/srccl/

11. M. Sangiacomo, 'It is time to cut use of plastics', *San Francisco Chronicle*, 24 December 2018, https://www.sfchronicle.com/opinion/openforum/article/It-is-time-to-cut-use-of-plastics-13489726.php

Fresh Water

—

Janet G. Hering

Let's close our eyes and think of the words 'fresh water'. What images come to mind? Perhaps we think of a gentle spring rain that nourishes wildflowers and newly planted crops. We may envision a clear lake with fish schooling just under the surface, or a rushing stream with migrating salmon leaping from the water. In wealthier countries or regions, we may think of filling a glass of water from the tap, or of our morning shower, often without realizing what a privilege it is to have safe water delivered reliably to our homes. We may also recall water being used to fight fires or for irrigation, transportation and hydropower. Negative images may also come to mind, such as the environmental devastation associated with water pollution or its diversion for agriculture or hydropower. Or we might think of the torrential downpours that cause flooding and the loss of human life and property. These examples illustrate that people across the globe can experience fresh water in vastly different ways.

Throughout history, water has always been of central importance for human welfare. Each person needs about 80 L of water per day for drinking, cooking and hygiene (30,000 L per year) and about a factor of forty more (1.2 million L per year) if dietary needs are included. Because of these essential requirements for water and the importance of natural waterbodies and watercourses for fishing, transportation and even defense, the geography

https://doi.org/10.11647/OBP.0193.25

of major rivers, deltas and coastlines has profoundly influenced the patterns of human settlement. Dating back to antiquity, great civilizations have also shaped the landscape of water features (or *waterscapes*) through diversions and damming of rivers, and the draining of wetlands. Some of these changes have left visible reminders, such as the ancient Roman aqueducts. But in the absence of such artifacts, we tend to forget how massive our influence has been. For example, few citizens of the United States realize that six States have lost over 85% of their wetlands. In Europe, it is impossible even to estimate such losses. In our modern world, the altered waterscape appears to us as natural.

Despite the successes of ancient civilizations in managing water supply and waste management, industrialization and urban growth in both Europe and North America were accompanied by severe outbreaks of water-borne disease. As described by David Sedlak in his book *Water 4.0: The Past, Present, and Future of the World's Most Vital Resource*, cholera and typhoid epidemics were common in the nineteenth and early twentieth centuries.[1] A major cholera outbreak in London in 1848 was famously traced to a well contaminated by sewage discharges to the River Thames. A series of improvements in water supply (including the protection of water sources, sand filtration and chlorination) led to massive improvements in public health in industrialized countries in the early- to mid-1900s.

At the time of the first Earth Day in 1970, the benefits of safe drinking water and improved public health were mainly taken for granted in wealthy countries. But these societies had failed to address the visible and extreme pollution of rivers and lakes resulting from the discharge of industrial wastes and inadequately treated sewage. Oil and industrial waste on the Cuyahoga River in Cleveland, Ohio, famously caught fire — not for the first time — in 1969. That same year, Lake Erie was pronounced 'dead' by major news agencies.[2] Massive fish kills resulted from both industrial pollution and introduction of nutrients (eutrophication) that created a 'dead zone' with insufficient oxygen for fish. These drastic environmental impacts motivated much of the activism of the first Earth Day as well as the establishment of the US Environmental Protection Agency (EPA). The US Clean Water Act of 1972 set the stage for massive improvements in environmental protection, including widespread upgrading of wastewater treatment. The impacts of industry on water quality were also a concern in Europe. In 1986, a fire at a chemical storage facility in Basel near the Swiss border shocked Europe with severe contamination of the Rhine River and a complete die-off of fish along the entire length of the river. As in the US, these events led to national and international legislation and agreements to curtail pollution and monitor water quality, as detailed in Frank Dunnivant's book *Environmental Success Stories: Solving Major Ecological Problems and Confronting Climate Change.*[3]

The first Earth Day did not need to focus on access to safe drinking water and adequate sanitation, since these problems had already been solved in industrialized countries. The situation was, however, quite different in low- and middle-income countries (LMICs). To improve access to safe drinking water and adequate sanitation, the United Nations adopted the Millennium Development Goals (MDGs) in 2000,[4] recognized a universal human right to water and sanitation in 2010, and adopted the Sustainable Development Goals (SDGs) in 2015.[5] As a result of these international commitments, between 2000 and 2017, the proportion of the global population using safely managed services increased from 28% to 45% for sanitation, and from 61% to 71% for drinking water. Even in the Least Developed Countries, access to safely managed drinking water increased from 25% to 35%.[6]

The SDGs challenge the nations of the world to address human development and the environment in a concerted manner. The theme of the 2019 UN Environment report — Global Environment Outlook (GEO) 6: Healthy Planet, Healthy People[7] — highlights our reliance on ecosystem services. The recognition of the vital importance of fresh water ecosystems, which are among the planet's most biodiverse habitats, has led some nations, notably Ecuador, India, South Africa and New Zealand, to formalize the rights of the environment to water by limiting diversions of water, requiring the maintenance of minimum flow regimes, or even by conferring the legal status of persons on rivers. Notwithstanding these developments, the integrity and function of fresh water ecosystems are subject to increasing pressures, which derive from population growth, urbanization, water pollution, unsustainable development and climate change. The GEO-6 report clearly concludes that the world is not on track to achieve the environmental SDGs.

Achieving the SDGs will require that we find balances between our direct uses of water and the maintenance of aquatic ecosystem function. Balanced solutions will vary depending on local contexts and national priorities. Ideally, any solutions would avoid unanticipated consequences of narrowly focused measures and take advantage of synergies among the different SDGs. For water, SDG 6 ('Ensure availability and sustainable management of water and sanitation for all') is an obvious focus for investment and action. But protecting and improving fresh water quality are also incorporated in the targets to reduce the environmental impacts of cities (SDG 11), reduce marine pollution from land-based activities (SDG 14) and to conserve and restore freshwater ecosystems and their use (SDG 15).[8] Thus, progress toward these targets would also help to achieve SDG 6.

We can take the successes in improving water quality in industrialized countries since the first Earth Day as examples of effective strategies and measures. But we must be aware that the successful strategies of water management developed in industrialized countries depended heavily on prior infrastructure investments, which also caused ecological harm. These successful strategies also reflected climatic conditions that are quite different from those in countries facing water insecurity today. In addition, many of today's water insecure countries are experiencing population growth and increasing urbanization and have not yet fully developed their industrial base.

One example that is both motivating and cautionary is the past success of industrialized countries in providing safe water supply and sanitation — the central objective of SDG 6. This success is the basis for today's conventional paradigm for water and wastewater management, which reflects the climate, geography and historical development of Western Europe and North America. In this paradigm, water is used to transport waste, and the rapid conveyance of storm water away from urban areas is prioritized to prevent flooding. Systems are highly centralized, and water supply and wastewater are strictly separated.

In water-scarce regions of LMICs, however, using water for waste conveyance is not a practical use of a scarce resource, and the required investment in sewers would be prohibitively expensive. An attractive alternative is to reclaim water, energy and nutrients from wastewater (or from fecal sludge). This approach, which relaxes the strict distinction between water and wastewater, is also being taken up by industrialized countries as they move toward circular economies and replace aging infrastructure. Looking to the future, it seems that the fully centralized model of industrialized countries may not be the dominant, much less the only, way forward.

Infrastructure investment provides another example where caution should be exercised in emulating the past successes of industrialized countries. Past infrastructure investments for water conveyance and storage increased water security and, through hydropower production, energy security. Today, LMICs face massive infrastructure deficits, which call for trillion-dollar investments.[9] Thousands of major hydropower dams, with capacities of at least 1 MW, are in the planning or construction phase with most sited in LMICs.[10] Future infrastructure construction and planning should take into account past experiences with the ecological and social consequences of such projects as well as projections regarding future climate that might compromise intended benefits. LMICs that heed the lessons learned regarding loss of ecosystem services of watercourses and the social and ecological impacts of land lost to reservoirs may choose to pursue alternative energy production through renewable wind and solar energy.[11]

Today, water pollution poses different challenges for industrialized countries and LMICs. Despite the past successes of industrialized countries in controlling industrial pollution and upgrading municipal wastewater treatment plants, not all problems have been solved. Some of the remaining challenges for these wealthy countries derive from the widespread use of pharmaceuticals and personal care products that are poorly removed in conventional wastewater treatment.[12] Micro-plastics also pass through wastewater treatment plants and are released into the environment with the effluent.[13] This problem is partly addressed by recent legislation prohibiting the use of micro-plastics in some consumer products, but plastic microfibers are released whenever synthetic fabrics are washed. Other challenges arise from diffuse pollution, which is most often associated with agriculture. Fertilizers and plant protection products (PPPs), including pesticides, herbicides and fungicides, are used in agriculture and can contaminate groundwater as well as inland and coastal waters.[14] PPPs are often found in groundwater and surface fresh water at concentrations that exceed risk levels for sensitive aquatic species. The runoff of agricultural fertilizers into coastal waters has resulted in the formation of oxygen-deficient dead zones extending over more than 10,000 km².

In some LMICs, point-source pollution is accompanying rapid industrial development. This is a particular concern with the expansion of chemical and pharmaceutical production in the Asia Pacific.[15] In rapidly growing economies, regulatory controls on the discharge of industrial effluents are often inadequately enforced. Import of single-use and poorly recyclable plastics from industrialized countries, ostensibly for recycling, has led to massive accumulation of plastics in both fresh waters and the ocean. As in industrialized countries, agriculture also accounts for much of the diffuse pollution in LMICs. Worldwide, both the highest and lowest uses of PPPs are in LMICs with the highest application rates of pesticides in Latin America and the lowest in sub-Saharan Africa. Fertilizer application is also lowest in sub-Saharan Africa resulting in depressed crop yields.

The diverse challenges related to water use across the globe call for locally-appropriate, sustainable balances between meeting direct human needs for water and preserving the capacity of the water environment to provide ecosystem services. Location is critical because water issues arise from local conditions (such as climate and past infrastructure investment) and viable solutions to water issues can only be implemented by people and institutions with local authority and responsibility.[16] At the same time, demands and pressures on water resources in a specific location often reflect consumption patterns in distant countries importing embedded (or virtual) water, especially in the form of agricultural products. In Switzerland, for example, over 80% of the national water footprint derives from imported goods and services.[17] The inherent linkages between water, food and energy pose challenges to conventional, sector-based approaches to resource management, but also offer opportunities to leverage synergies. Water use in agriculture, for example, could be reduced by application of sensors and precision technologies to avoid excessive irrigation, re-use of treated wastewater and adoption of diets that include less meat.

Progressing toward a circular economy will be a key element in decoupling human well-being from resource exploitation, including unsustainable demands on fresh water.[18] Recovery of the nutrients nitrogen and phosphorus from human excreta could provide renewable and less energy-intensive fertilizers for agriculture. Separated plumbing systems for greywater (i.e., wastewater not contaminated with fecal waste) could allow in-home re-use, substantially decreasing the demand on the water supply to the household. Decentralized systems for water and wastewater could reduce the need for water distribution systems and sewers with their associated costs and often-substantial loss of water through leaks.

There will not be a 'one-size-fits-all' solution to meet the water-related challenges of the SDGs. We will need to work cooperatively to develop a shared portfolio of approaches that can be adapted for local conditions.[19] Cooperation must extend across sectors and include the co-production of knowledge by actors with different expertise, backgrounds, experience and responsibility. Advances in technology must be effectively harnessed, including new sensors, data technologies and real-time process control that can dramatically improve water efficiency. Open access to data will be critical to maximizing water efficiency and reducing environmental harm. At the same time, data platforms can raise public awareness of the importance and vulnerability of fresh water systems and encourage citizen engagement. More generally, open access to scientific knowledge could promote evidence-based and participatory water management.[20]

To realize the SDGs, we can take the attributes of water as a guide. Water flows around obstacles — SDG implementation must also be adaptive and appropriate to local contexts. Over time, water can wear away the hardest stone — we must also be persistent. Water has great power that can be destructive but can also be harnessed productively — as individuals and collectively, we too have the power to transform our societies. The Sustainable Development Goals demand nothing less.

Endnotes

1. D. Sedlak, *Water 4.0: The Past, Present, and Future of the World's Most Vital Resource*, New Haven: Yale University Press, 2014, 41–62.

2. See 'Oceans 2020' by David M. Karl in this volume.

3. F. Dunnivant, *Environmental Success Stories: Solving Major Ecological Problems and Confronting Climate Change*, New York: Columbia University Press, 2017, 129–39, https://doi.org/10.7312/dunn17918

4.   Available at https://www.who.int/topics/millennium_development_goals/about/en/

5.   Available at https://sustainabledevelopment.un.org/?menu=1300

6.   UNICEF and WHO, *Progress on Household Drinking Water, Sanitation and Hygiene 2000–2017: Special Focus on Inequalities*, ed. R. Steele, New York: UNICEF and WHO, 2019, https://washdata.org/sites/default/files/documents/reports/2019-07/jmp-2019-wash-households.pdf

7.   UNEP, *Global Environment Outlook — GEO-6: Healthy Planet, Healthy People*, ed. P. Ekins, J. Gupta and P. Boileau, Nairobi: UNEP, 2019, https://doi.org/10.1017/9781108627146, https://www.unenvironment.org/resources/global-environment-outlook-6

8.   J. Sachs, G. Schmidt-Trabu, C. Kroll, G. Lafortune and G. Fuller, *Sustainable Development Report 2019: Transformations to Achieve the Sustainable Development Goals*, New York: Bertelsmann Stiftung & Sustainable Development Solutions Network, 2019, https://www.bertelsmann-stiftung.de/en/publications/publication/did/sustainable-development-report-2019/

9.   The World Bank, *Rebalancing, Growth, and Development: An Interconnected Agenda*, Washington, DC: The World Bank, 2011, https://siteresources.worldbank.org/INTWDRS/Resources/WDR2011_Full_Text.pdf

10.   C. Zarfl, A. E. Lumsdon, J. Berlekamp, L. Tydecks and K. Tockner, 'A global boom in hydropower dam construction', *Aquatic Sciences*, 2015, 77, 161–70, https://doi.org/10.1007/s00027-014-0377-0

11.   J. Opperman, J. Hartmann, M. Lambrides, J. P. Carvallo, E. Chapin, S. Baruch-Mordo, B. Eyler, M. Goichot, J. Harou, J. Hepp et al., *Connected and Flowing: A Renewable Future for Rivers, Climate and People*, Washington, DC: WWF and The Nature Conservancy, 2019, http://d2ouvy59p0dg6k.cloudfront.net/downloads/connected_and_flowing__wwf_tnc_report__5.pdf

12.   See also 'The Global Chemical Experiment' by Elsie Sunderland and Charlotte C. Wagner in this volume.

13.   See also 'Earth and Plastic' by Roland Geyer in this volume.

14.   J. Mateo-Sagasta, S. M. Zadeh and H. Turral (eds.), *More People, More Food, Worse Water? A Global Review of Water Pollution from Agriculture*, Rome: FAO and IMWI, 2018, http://www.fao.org/3/ca0146en/CA0146EN.pdf

15.   UNEP, *GEO-6 for Industry in Asia-Pacific*, Nairobi: UNEP, 2019, https://www.unenvironment.org/resources/report/global-environment-outlook-6-industry-asia-pacific

16.   J. G. Hering, D. Sedlak, C. Tortajada, A. K. Biswas, C. Niwagaba and T. Breu, 'Local perspectives on water', *Science*, 2015, 349, 479–80, https://doi.org/10.1126/science.aac5902

17. WWF, *The Swiss Water Footprint Report*, Switzerland: WWF, 2012, https://www.eda.admin.ch/dam/deza/en/documents/publikationen/Diverses/209748-wasser-fussabdruck-schweiz_EN.pdf

18. T. A. Larsen, S. Hoffmann, C. Luthi, B. Truffer and M. Maurer, 'Emerging solutions to the water challenges of an urbanizing world', *Science*, 2016, 352, 928–33, https://doi.org/10.1126/science.aad8641

19. J. G. Hering, 'Water: The environmental, technological and societal complexity of a simple substance', in *Encyclopedia of Water: Science, Technology, and Society*, ed. P. Maurice, New York: John Wiley & Sons, 2019, 1–9, https://doi.org/10.1002/9781119300762.wsts0038

20. WWF-US, *Free Flowing Rivers*, http://freeflowingriver.org/about; Earthwatch Institute, *FreshWaterWatch*, https://freshwaterwatch.thewaterhub.org/content/citizen-science

Media

—

Candis Callison

Two years prior to the first Earth Day, on December 24, 1968, the Apollo 8 mission returned with *Earthrise,* the first color image of our pale blue planet. Taken by US astronaut Bill Anders, the image appeared at a time when the Vietnam War, the Cold War and other geopolitical entanglements dominated news around the world. Many now look at *Earthrise* as the first in a series of galvanizing images for those concerned about the environment. A similarly iconic image, *The Blue Marble*, came from the Apollo 17 mission in 1972, the last manned flight to the moon. This image encompasses the whole Earth, showing the Southern polar ice cap for the first time, most of the African coastline, and some of Asia. Even though heavy clouds shroud many parts of it, *The Blue Marble* has become one of the most recognizable images of our shared planet, thanks in no small part to the role played by many kinds of media in spreading this image around the world.

These images of Earth suspended in its solar system have long been credited with a shift in our conceptualization of the planet, giving primacy to environmental borders instead of political ones.[1] They also contributed to a sense of the planet as vulnerable and at risk. Two decades after NASA distributed the Apollo images, *Time* magazine put 'Endangered Earth' on its 1989 cover as 'Planet of the Year', instead of its usual 'Person of the Year'.[2] The previous year, NASA scientist James Hansen had presented the first testimony about climate change to the US Senate. And the year before that, in 1987, the Montreal Protocol had been signed, banning the global production of chlorofluorocarbons (CFCs), chemicals that contributed

https://doi.org/10.11647/OBP.0193.26

to the Antarctic ozone hole.[3] By 1990, the US was passing strong new amendments to the Clean Air Act,[4] the Intergovernmental Panel on Climate Change (IPCC) was releasing its first Assessment, Carl Sagan had joined with thirty-two other Nobel Laureates to issue a letter titled 'Preserving and Cherishing the Earth: An Appeal for Joint Commitment in Science & Religion', and Pope John Paul II called for environmental responsibility on the World Day of Peace. These heady and hopeful times of environmental awareness were well covered by dominant media at the time: television, radio, newspapers.

The 1990s, however, gave way to shifts in media, science and environmental movements. While many point to the energizing social and political impact of the 1992 Earth Summit in Rio de Janeiro, still dominant legacy media began to shift their gazes away from environmental issues and science coverage more generally. Some have speculated that entertainment and celebrity news took over, displacing environment, science and other kinds of stories. What's clear looking back is that Earth lost some of its celebrity status, while, at the same time, science coverage began to decrease during this decade. Moreover, by the time the 1997 Kyoto Protocol was signed by member nations of the United Nations Framework Convention on Climate Change (UNFCC), questioning the very notion of climate change had already become entrenched in some national media contexts.[5] Successive IPCC reports in 1995 and 2001 were progressively more clear, urgent, and confident in their conclusions, but media coverage still varied across different countries. This became particularly problematic in the US, as news stories on climate science were likely to include 'both sides', giving equal weight to climate change deniers despite strong and continuing scientific consensus. Support for the Kyoto Protocol waned — the US pulled out under President George W. Bush Jr.; public interest and engagement with climate change reached an all-time low. Many scientists, activists and policymakers blamed the media for not doing their job, and some journalists accepted the blame. But there were also other complex problems at play, and some of these still inform today's media in their coverage of climate change.

It might seem surprising to the new generation of 'climate-striking' youth, but climate change, in its earliest years, was not considered a timely issue, nor one with a well-defined range of possible and likely ramifications. As many reporters have pointed out to me over the last twenty years, climate change is a difficult story to tell — in part because it 'oozes' and doesn't 'break'. In other words, incremental stories on findings related to aspects of climate science generally do not conform to the 'news values' of timeliness, prominence and immediate impact — that is, unless climate change can be pegged to an event or famous person. IPCC scientists winning a Nobel Prize, along with Al Gore, or Greta Thunberg's activist speeches spring to mind as good examples. These kinds of stories produce awareness, but whether or not they serve to educate or engage the public about climate change depends on *how* these stories are told.

Science education also poses some problems for reporting on climate science. Many journalists have told me that they aim at an early high school education level of science knowledge — corresponding to somewhere between US ninth and eleventh grade (ages fourteen to sixteen).[6] Public education is thus always part of the task embedded in any story touching on climate change. Those who reported on it in the 1990s were cautious and careful in their efforts at public education. But, by the early 2000s, even as the science affirmed and confirmed earlier assertions, journalists who reported on climate change as an urgent issue were accused of being alarmist, unbalanced, or biased in their coverage. On the other hand, those journalists and news organizations who gave it scant coverage were equally castigated by scientists and activists for not being effective in their reporting, thus failing to mobilize public and political engagement.

It wasn't until the mid-2000s that attention to climate change from news media began to shift, in part due to the extreme weather events. Perhaps more than any other weather event, hurricanes provide a case study on how a future with climate change can be made topical — and frightening. In the aftermath of the devastating 2017 hurricane season, climate scientist Michael Mann described the relationship between climate change and Hurricane Harvey in an op-ed for *The Guardian,* stating that: 'While we cannot say climate change "caused" Hurricane Harvey (that is an ill-posed question), we can say that it exacerbated several characteristics of the storm in a way that greatly increased the risk of damage and loss of life. Climate change worsened the impact of Hurricane Harvey'.[7]

Even for Mann — a scientist used to the media spotlight and experienced in speaking to diverse audiences — the nuance required to characterize causal factors and links to climate change is a challenge, involving a reframing of what's in question. Like most climate-related findings, understanding the relationship between hurricanes and climate change requires a certain amount of knowledge about the difference between *climate* and *weather* — and, how scientific facts are formed, evolve and build on one another. Hurricanes Katrina and Rita in 2006 were a kind of precursor to this, as journalists and activists (think back to Al Gore's *An Inconvenient Truth* film poster) rushed to make causal links even while the science related to hurricanes and the risks of coastal development were much more nuanced and complex.

Legacy media — the same media that sent *Earthrise* and *The Blue Marble* photos around the world — have traditionally been tasked with calling the widest possible audiences to attention on issues of immense importance. Yet, legacy media are no longer able to rally large swaths of the public in ways previously expected. Many media are struggling with changing audience habits, faltering economic models and labor precarity, resulting in decreased resources for newsrooms and declining numbers of journalists able to earn a living wage. Science journalists have become as rare as the speciality knowledge and interest that comes with working on 'a beat'. Only the prestige press — those few news organizations with healthy subscription numbers *and* benefactors (e.g., *The Washington Post* and *The New York Times*) — or those who deem it a priority area based on their market or mandate (e.g., public broadcasters) have the luxury of employing reporters able to work in a single subject area over a long period of time.

Even more pressing than changes to legacy media are the changes wrought by social media and the sheer volume of information available on a daily basis. Social media now plays a huge role in what news gets read and shared, and the impact it has on what stories news organizations choose to cover is still unclear. For example, will social media uptake of climate news stories contribute to a shift in news values and editorial interests, so that climate change becomes a prominent topic worthy of consistent attention beyond clearly linked events, reports or disasters? At the moment, it is too early to tell.

Investing an issue like climate change with meaning, ethics and morality has generally been the work of social groups — conversely, this implies that one's social group can also undermine confidence in scientific facts and the immediacy of concerns about climate change.[8] Scholars who study media, science and social movements have been arguing for some time that diverse communities play a huge role in the circulation of facts, but the exact and varied nature of this influence is only just beginning to be understood. This is a departure from the view that audiences demand and require textbook-like scientific facts. Instead, the work of journalism might better be considered as helping audiences develop a relationship with evolving climate facts that have been occasionally revised, but more often elaborated and affirmed.

When social movements are active online — providing source materials, perspectives and reporting — it becomes even more essential that journalists understand the varied contexts and audiences for their reporting, whilst also introducing new ideas about accountability into the discussion. The *Columbia Journalism Review* recently argued that climate change connections should be articulated when reporting on all environmental concerns.[9] While it makes good 'news' sense to link high profile disasters like hurricanes and fires, it is equally important for business and policy decisions to be linked to climate change. Holding corporations and governments accountable to ever clearer metrics for climate mitigation and adaptation may not be an easy news story to pitch or tell, but, in the longer term, it is an arguably more imperative one.

Perhaps an even greater challenge to address in media coverage is the representation of diversity in human experience and relations, and the importance of acknowledging the long histories of these relations. Erasing ongoing and historic relations between humans and non-humans, or even recognizing the inter-connectedness of our social structures and societies doesn't just wave away the challenges of contending with them, even in the name of shared concerns. Consider, for example, how Indigenous communities have often either fallen out of coverage of climate change, or been relegated to the role of victims and occasionally heroes. In general, media have not done well in reporting on issues related to race, Indigeneity and colonialism, and this is particularly true in the context of climate change.[10] Yet, Indigenous peoples comprise 5% of the world's population and live in over ninety countries — and their communities, particularly those in the Arctic, sub-Arctic and southern Island nations, are likely to suffer disproportionately from the impact of climate change.

Indigenous communities offer deep oral histories, distinctive knowledge, instructive frameworks for relations with non-humans and humans, and pathways towards resilience in the face of enormous changes. Many diverse Indigenous communities tell stories not only of the immediate environmental changes they experience, but how these changes layer onto long histories of their lands, relations with colonial powers and the imposition of capitalism.[11] Understanding what climate change means and mandates in these varied global and Indigenous contexts forces us to examine deeply embedded questions about how we relate to each other, how our institutions and policies reflect these relations and how we might move toward a more just world even while contending with unequal climate disruptions.

Our myriad of handheld devices now connect us globally in ways that weren't imaginable fifty, or even twenty, years ago. In the last couple decades, our daily lives have been radically transformed by the influx of media and information. For many of us, regardless of where we live or our socio-economic status, mobile devices have become an essential conduit for navigating some (or many) aspects of daily life. Yet, with so much digital infrastructure, how connected we feel to each other and how informed we are is still very much an open question — and one that has consequences for global issues like climate change, with all its attendant potential and probable future impacts.

We have watched as, in the last decade, activists young and old have utilized a variety of media to call for action in the face of climate change. Still, news media continue to play an outsized and shifting role even as digital infrastructures produce more information sources and make it easier to educate and mobilize some publics and policymakers on climate change. How news media navigate their own structures, crises and persistent challenges in telling the story of climate change will be critical as more data, catastrophes and risks related to climate change become apparent. And while our social lives will continue to play an important role in what media we choose to follow, it will also become increasingly vital to understand what images like *The Blue Marble* obscure: the histories of our relations, and the systems that inform our understanding of how to live together in a shared future with climate change.

Endnotes

1. Yet, as scholars have continually pointed out, efforts at erasing existing national and political differences in an attempt to offer common solutions still often replicate familiar social and economic divisions and power hierarchies. See for example, S. Jasanoff, 'Heaven and earth: The politics of environmental images', in *Earthly Politics*, ed. S. Jasanoff and M. Long Martello, Cambridge, MA: MIT Press, 2004, 31–52.

2. *Time* magazine's person of the year for 2020 is seventeen-year-old Swedish climate activist Greta Thunberg.

3. Available at https://ozone.unep.org/treaties/montreal-protocol-substances-deplete-ozone-layer/text

4. See also 'Air' by Jon Abbatt in this volume.

5. Available at https://unfccc.int/resource/docs/convkp/kpeng.pdf See also M. Boykoff, *Who Speaks for the Climate? Making Sense of Media Reporting on Climate Change*, Cambridge, UK: Cambridge University Press, 2011, https://doi.org/10.1017/cbo9780511978586

6. C. Callison, *How Climate Change Comes to Matter: The Communal Life of Facts*, Durham, NC: Duke University Press, 2014, https://doi.org/10.1215/9780822376064

7. M. E. Mann, 'It's a fact: climate change made Hurricane Harvey more deadly', *The Guardian*, 28 August 2017, https://www.theguardian.com/commentisfree/2017/aug/28/climate-change-hurricane-harvey-more-deadly

8. Callison, *How Climate Change Comes to Matter*, 2014.

9. J. Allsop, 'Climate change plays second fiddle as California burns', *Columbia Journalism Review*, 12 November 2018, https://www.cjr.org/the_media_today/california_wildfires_climate_change.php. See also USGCRP, *Impacts, Risks, and Adaptation in the United States: Fourth National Climate Assessment, Volume II: Report-in-Brief*, Washington, DC: US Global Change Research Program, 2018, https://nca2018.globalchange.gov/downloads/NCA4_Report-in-Brief.pdf

10. C. Callison and M. Lynn Young, *Reckoning: Journalism's Limits and Possibilities*, Oxford and New York: Oxford University Press, 2020, https://doi.org/10.1093/oso/9780190067076.001.0001

11. See C. Callison, 'Climate change communication and Indigenous publics', in *Oxford Research Encyclopedia of Climate Science, Vol. I*, ed. M. Nisbet, Oxford: Oxford University Press, 2017, 112–32, https://doi.org/10.1093/acrefore/9780190228620.013.411 and K. P. Whyte, 'Justice forward: Tribes, climate adaptation and responsibility', *Climatic Change*, 2013, 120, 517–30, https://doi.org/10.1007/978-3-319-05266-3_2, https://kylewhyte.marcom.cal.msu.edu/wp-content/uploads/sites/12/2018/07/Justice_Forward_Tribes_Climate_Adaptatio.pdf

Space Junk

—

Alice Gorman

In a famous scene from the 2008 animated movie *WALL-E*, a rocket streaks up from Earth's surface, a tiny robot clinging to its side. Before reaching open space, the rocket punctures a thick layer of old satellites jostling cheek by jowl. As the layer cracks open, space junk erupts, scattering satellites like winged flies. A silver flagellated sphere gets caught in the robot's head before drifting off. The angles of its antenna are instantly recognizable; it's Sputnik 1, the first satellite ever launched, now reduced to just another piece of junk in the orbiting scrap yard. The message from this future world is clear. Just as Earth's surface has been polluted to the point where all human life has abandoned it, so too has the space surrounding Earth been choked with the endlessly circulating junk of the late industrial age.

Once upon a time, humans and their ancestors looked into the night sky to see the light of the Moon, stars, planets and galaxies, weaving them into culture through science and stories. Celestial bodies were intimate partners in the creation of a cosmos inhabited by ancestral beings and living impulses. Starting from the fifteenth century, however, new scientific methods like the telescope began to transform this intimate landscape into an infinite universe where human concerns were irrelevant.[1] This process created a distance between us and the stars that we have been striving to close ever since.

In the twentieth century, the means of creating intimacy with sterile space became material rather than visual. On October 4, 1957, the Russian satellite Sputnik 1 was propelled

     https://doi.org/10.11647/OBP.0193.27

into Earth orbit. It was barely visible to the naked eye, but it made people look upwards in wonder or in fear. Sputnik 1 was blind. It carried no cameras to image the Earth whirling beneath it, but it did speak, emitting a beeping radio signal at the frequency of 20 MHz that became the sound of the Space Age. For four weeks (until the November 3 launch of Sputnik 2 with Laika the dog on board), Sputnik 1 was the only human object beyond Earth. Its successful injection into orbit was a moment of enormous consequence. It transformed Earth orbit into a buffer zone between humans and the wider solar system. In the years that followed, the formerly featureless 'orbital space' rapidly accumulated a population of robotic satellites and the junk they generated in their decay.

In the orbital space surrounding Earth, objects are in continual movement, and places are defined by velocity and height above the planet's surface. This is no longer a geography, which maps places on Earth, but an orbitography. Over the past six decades, human objects have colonized this orbital space, dividing it into zones and regions with distinct characteristics.

Low Earth orbit (LEO) ranges from around 200 km to 2,000 km above Earth's surface. Within this range, Earth observation satellites provide daily meteorological observations, environmental monitoring and military surveillance. Spacecraft in LEO are still within the outer reaches of Earth's atmosphere, which means their presence is temporary. Sparse molecules of gas exert friction on the objects, slowing them down and lowering their orbit until eventually they are drawn into the upper layers of the thermosphere. Few objects survive the ensuing blaze of re-entry into the atmosphere. Those which do make it through intact tend to have very high melting points, like the spherical titanium pressure vessels from spacecraft propulsion systems. Often, these 'space balls' are found on Earth's surface years after their re-entry, lying forgotten in fields or by lake shores.

In the first six years of the Space Age, all spacecraft were launched into LEO. The oldest space artefact still in existence is the Vanguard 1 satellite, launched by the United States in 1958. Like Sputnik 1, this satellite was also a polished silver sphere, but with six antennas as opposed to Sputnik's four. Notably, Vanguard 1 carried the first solar panels in space, providing energy to power its mission. Its orbit, tracked eagerly by people watching with binoculars and telescopes on Earth, was not a smooth curve. Wobbles in its trajectory caused by variations in Earth's gravitational pull demonstrated that Earth is not a perfect sphere; its oceans and continents hide a lumpy surface underneath. Both Sputnik 1 and Vanguard 1 were launched as part of the 1957–1958 International Geophysical Year. These early spaced-based observations proved to be a watershed in understanding the Earth system.

Earth orbit is a machine landscape, as human bodies are not adapted to the hostile conditions of the space environment.[2] Despite this, the most famous inhabitant of LEO is the International Space Station (ISS), which was launched in 1998 into an orbit around 400 km above Earth's surface. The ISS, weighing 420 metric tons, has been continuously occupied for twenty years, and is, by far, the largest artificial object in orbit, representing about 5% by mass of all human-made space materials. Inside its metal tubes, a crew of two to six people have lived in weightlessness from a few days to over a year at a time. The primary purpose of the ISS is to carry out science in microgravity, but its presence makes Earth orbit a home, a place where a new culture is being created through shared experiences of life in space.

Beyond LEO, starting at around 2,000 km above Earth's surface, is medium Earth orbit (MEO). In this region, high energy charged particles streaming from the Sun are trapped in Earth's magnetic field, enclosing the planet in protective flower-like curved petals. The high radiation levels are dangerous for satellites, potentially damaging their delicate electronics. Nonetheless, navigation satellites, vital to many facets of our everyday lives, are located in this region. The US Global Positioning Satellite (GPS), the European Space Agency Galileo and the Russian GLONASS constellations all orbit at around 20,000 km in MEO. Other MEO satellites include the 1962 commercial telecommunications satellite, Telstar 1, which inspired a raft of popular culture responses, including the design of a black-and-white hexagonal soccer ball and a chart-topping pop music hit (*'Telstar'* by the Tornados). When Telstar 1 failed in 1963, it became another piece of space junk, but one with great cultural significance.

High Earth orbit (HEO) begins about 35,000 km above Earth's surface. This is the region where telecommunications satellites, as well as the Chinese BeiDou constellation of navigation satellites, are located. At this altitude, satellites in geostationary orbits travel at the same speed as the rotation of Earth, maintaining a fixed position above a particular point on the planet's surface. The science fiction writer Arthur C. Clarke,[3] drawing on the work of the early twentieth century space theorists Konstantin Tsiolkovsky and Hermann Noordung, was the first English language writer to describe the potential of these orbits. In 1945, Clarke proposed that just three satellites in geostationary orbit could provide coverage of the entire globe. About 500 km above this orbit is a graveyard where old telecom satellites are boosted out of the way of functioning spacecraft. HEO ends where cislunar space begins, at around 150,000 km from Earth. Some spacecraft have passed through this region, like the STEREO A and B solar observatories, and it's possible that dust from satellite decay has migrated here.

For sixty years, we have been adding human materials to the space environment. These spacecraft have transformed virtually every aspect of our lives — from agriculture, environmental management and weather prediction to internet and banking. But what happens when their official mission ends? Suddenly, their status changes from an asset to a liability. They become 'space junk'.

In the first decade of the Space Age, scientists were concerned about the dangers of meteorites colliding with astronauts and spacecraft. It became apparent, however, that human debris was coming to outnumber the 'natural' objects in orbit. Instead of micrometeorites, the real problem was likely to be collisions between human materials. In 1978, Donald Kessler and Burton Cour-Palais wrote a paper which predicted a worst-case scenario, now known as the Kessler syndrome.[4] Continued debris collisions, they argued, could result in a runaway cascade where debris would be created even if no new objects were launched. In this scenario, certain regions of space could effectively become unusable, as depicted in *WALL-E*.

Today, there is significant debate about how close we are to realizing the Kessler syndrome. But there is no doubt that the risk of collision with space junk is increasing. In 1970, the year of the first Earth Day, there were an estimated 2,500 space objects distributed from LEO to HEO. Half a century later, in 2020, there are well over 30,000 pieces of debris larger than 10 cm in Earth orbit, and many millions of fragments and particles below that size. The combined total weight of human-derived space junk is estimated to be 8,400 tons (the equivalent of 4,000 adult giraffes). This includes functioning satellites, whole satellites that are no longer working, rocket bodies left abandoned in orbit after delivering their payloads, mission-related debris like the fairings that are discarded to release the satellite within, and chunks, fragments and flecks of spacecraft materials.

The density of junk is greatest in LEO. And although objects in LEO eventually get dragged back into the atmosphere where they largely disintegrate, this removal happens at a much slower rate than the creation of new debris. Over the past several decades, the space debris population has increased dramatically, as the global economy and everyday life has come to depend more and more on satellite technologies. Sometimes, catastrophic events cause a sudden increase in the amount of debris. This was the case in 2007 and 2019, for example, when China and India deliberately destroyed their own satellites using Earth-launched missiles, leading to the creation of thousands of debris objects in LEO. These actions were widely condemned by the international space community, but there's no guarantee similar tests won't occur again.

Like the accumulation of plastics on Earth, the growth of space junk poses significant problems. Satellites are a billion-dollar industry, upon which much of our modern lives depend. Collision with space junk can erode a satellite's surfaces, cause it to malfunction, or, in the worst-case scenario, explode. Each collision creates new pieces of space debris, further exacerbating the problem. The risks of space junk could prevent the emergence of the much-anticipated space tourism industry.

To date, solutions to our growing space junk problem have included guidelines to minimize the creation of debris. These guidelines recommend designing spacecraft so that there is no explosive fuel left at the end of mission life; removing spacecraft to a 'graveyard orbit'; shielding spacecraft against collision and incorporating tethers to drag them into the atmosphere. As for actively removing old debris from orbit — something that is now actively planned[5] — there are two main obstacles. First, maneuvering in orbit to capture an old satellite is extremely costly in fuel, and therefore presents a poor business case, even for the most potentially dangerous objects. More importantly, any mechanism for removing satellites from orbit could be deployed as a weapon to hobble an adversary's space capabilities, creating a host of geopolitical challenges. And so, despite increasing attention being given to space situational awareness (SSA) and space traffic management (STM), we have thus far made little progress in solving the growing problem of space junk. Time, however, may be running out.

At the same time, it may be too easy to characterize space debris as merely a problem of 'junk' that needs to be fixed. There are other ways of understanding what Earth orbit has become. One alternative approach is to consider Earth's near space environment as a cultural landscape with its own intrinsic values. When viewed through such a lens, we come to break down the distinction between natural and cultural, envisioning a new space that has resulted from the historic interactions between human and environmental factors. Here, interplanetary dust mingles with the machine dust derived from the decay of human-manufactured materials under the harsh conditions of high energy particles, micrometeorites, atomic elements and collision with other space debris. This dust mix is the archaeological signature of a space-faring species.

What counts as 'junk' is also very dependent on cultural values. Among the 4,000 defunct satellites in Earth orbit, many have heritage value in preserving legacy technologies, historic moments or processes, or through their symbolic or social significance to a nation or community. The natural setting for these artefacts is the orbital landscape, and where they do not constitute a collision risk, there is no reason to remove them. Moreover, old satellites or satellite materials can be recycled or re-used. Abandoned satellites can be repurposed for new missions such as collecting scientific data, providing they have sufficient fuel or batteries left. The metals used in spacecraft manufacture can also be used as fuel in plasma rocket engines. In future orbital manufacturing industries, space scavenging could save the enormous expense of lofting materials from Earth. Clearly, end-of-life plans for satellites have thus far not been creative enough.

Today, in 2020, we are facing a transformation of Earth's orbital landscape with the launch of proposed mega-constellations of internet telecommunications satellites. The first of these have already been launched, even though the effects of injecting tens of thousands of new objects into an already congested region of space are not fully understood. Notwithstanding the optimistic assurances of commercial operators that the satellites will quickly re-enter Earth's atmosphere, it is clear that that predictions of the onset of the Kessler syndrome will have to be revised.

No longer will people on Earth have to scan the skies systematically to pick out a lone silver sphere, as they did in 1957. Satellites sightings will become the norm, rather than the exception; they will be our constant companions whenever we look heavenwards. The burning shards of re-entering spacecraft will cease to cause fear and astonishment. And, in a few decades, the people who remember the sky before Sputnik 1 shattered its peace will be gone. Soon, the whirling graveyard of space junk punctuated by living robots will be all we have ever known.

Endnotes

1.   A. Koyré, From Closed World to Infinite Universe, Baltimore: Johns Hopkins University Press, 1957.

2.   A. C. Gorman, 'Ghosts in the machine: space junk and the future of Earth orbit', Architectural Design, 2019, 89, 106–11, https://doi.org/10.1002/ad.2397

3.   A. C. Clarke, 'Extra-terrestrial relays', Wireless World, 1945, 305–08.

4.   D. Kessler and B. Cour-Palais, 'Collision frequency of artificial satellites: The creation of a debris belt', Journal of Geophysical Research, 1978, 83, 2637–46, https://doi.org/10.1029/JA083iA06p02637

5.   A new ESA mission called ClearSpace-1, with the aim of piloting space junk removal, is planned for launch in 2025. See http://www.esa.int/ESA_Multimedia/Images/2019/12/ClearSpace-1

Saving the Boat

—

Zoe Craig-Sparrow and Grace Nosek

I, Zoe Craig-Sparrow, was born and raised on the Musqueam reserve in British Columbia, and grew up fishing salmon on the Fraser River with my family. The currents are swift here, and the water is dangerous. Many, including members of my family, have lost their lives on the River.

In my childhood and teens, I served as a deckhand under my grandfather (an experienced commercial fisherman) and my mother (who was also raised fishing on the water), soaking up their knowledge and experience, accumulated over many generations.

When I was twenty, I got my first chance to captain my own boat on the River, accompanied by a deckhand and my thirteen-year-old sister, Charlee. We were fishing along a bend of the river, in a place with notoriously swift currents and many rocks. At one point, the currents became too strong, tearing our net and pushing it into our engine, where it soon became hopelessly tangled. I screamed to cut the engine, which we did, but it was too late. The propellers were wrapped up in the fragments of the net and the engine was useless; we were sitting ducks. The current immediately began to pull us dangerously close to the rocks, which threatened to puncture our fiberglass boat and sink us. I knew that we were in mortal danger, and that the lives of the boat's occupants were in my hands.

        https://doi.org/10.11647/OBP.0193.28

In the physical scramble to save the net and the boat, I was burned and cut by rope, and (to make matters worse) also stung by a wasp. My hand immediately started swelling and burning, but I had no time to register the pain. I felt sick to my stomach, the adrenaline pumping through my veins.

I had never dropped the anchor before, and was running through my training in my head (how much rope needs to be dropped, where, and how). I soon came to another terrible realization; the anchor was not attached to the boat. With my burned and swollen hands, I scrambled to secure the anchor to the boat, using the knots that I'd learned as a young girl. In those precious seconds, which I imagine now must have seemed like an eternity, I managed to successfully drop the anchor, as the deckhand pulled the snag free and brought the now useless net into the boat. We were momentarily out of mortal peril, but we now needed to keep the boat away from the rocks until help arrived. Getting a rush of inspiration, I grabbed the hook we used to pick up the net, which was attached to a long wooden stick. My sister, Charlee, used the stick to push us away from the rocks until help arrived. After a tense period of waiting, my family came to tow us back to safety.

Today, we (Zoe and Grace) are grappling with the great peril our world and our future are facing. Our rapidly warming world is pushing already strained social and ecological systems to the brink, profoundly threatening humans and animals alike. We know that climate change is already devastating vulnerable and marginalized communities — including Indigenous peoples, racialized minorities, women and people in the Global South, among others — and that such human devastation will increase exponentially if we continue on our current path. We are in the boat, engine destroyed, being dragged dangerously close to the rocks.

Our course is made more difficult by the powerful societal currents that are pushing our boat straight into the rocks. For decades, the fossil fuel industry and its allies have spent hundreds of millions of dollars undermining climate science and action, leaving many people unaware of the true dangers we face.[1] Companies like ExxonMobil have sowed doubt about whether climate change is real, serious, human-caused, and even whether it can be solved by humans.[2] They have profited, and protected their own infrastructure from climate change,[3] while making the rest of us feel helpless. Now, they are investing heavily in defeating climate legislation around the world while holding themselves out as climate leaders.[4]

Governments have been slow to address the climate threat, in no small part because of the fossil fuel industry's coordinated efforts to undermine climate action. Voters have not always signalled robust support for bold climate action, and politicians have not demonstrated the political courage to act ahead of public opinion or anger entrenched corporate interests.

Faced with these institutional failures, it may feel impossibly daunting to engage with the threats of climate change, biodiversity loss and other environmental crises.

But think of Zoe, with her hands burned from the rope and swollen with wasp venom, far from help and terrified for the safety of her little sister. Think of Zoe finding the courage, strength and creativity to save her boat and its occupants against all odds.

Think of the courage of Indigenous land defenders putting their bodies on the line, time and time again, to protect the land, water and air for all of us in the face of colonial and corporate forces often prepared to use lethal violence against them.[5] Think of the youth climate strikers around the world, some still in elementary school, recognizing that justice for the planet requires justice for all humans, and leaving their childhood joys behind to demand systemic social, political and economic change.

Like Zoe's ingenious scramble for a tool to keep the boat off the rocks, these groups are coming up with creative solutions to the climate crisis. They are centering long-ignored voices to dream up new futures. Futures where we conserve energy by working fewer hours and consuming less, where we have more time to connect, play, build community and pursue our passions. Futures where every human born in this world has the right and the opportunity to thrive, and where we do not mercilessly exploit non-human animals.

It will often seem easier to tune out, to hope that someone else will take up the task of saving the boat. We are all busy in the twenty-first century, and many of us are locked into the current economic and political system — worried about job security, paying the mortgage or continuing to afford the same quality of life. It can be scary to work for systemic climate action, to call or write your represented officials, to speak to friends, family and strangers about electing climate justice advocates. In a world where there is no pause, it is easy to feel overworked, overwhelmed and too tired to engage.

But know this — young people, often led by Indigenous youth, recognize that we're on the boat. We're reeling and sick, viscerally aware of the immense suffering caused by inaction on climate change. We can see the rocks ahead and it terrifies us. And yet, we are showing up every single day to try and guide the boat away from danger, knowing that it is the most vulnerable who are already in the water. We see the danger, but we also know where and how to guide the boat to safer, more just, more intersectional waters. We're pulling at the rope, grasping at the anchor, getting buffeted and knocked down in the process. And yet, we are still showing up in the streets and in the institutions of power, week after week. We need your help. We need it now. Help us tell the story of the true dangers we face, of the vulnerable people who are already suffering, to whatever community you are a part of — your reading group, faith organization, recreational sports team, alumni group, place of work, financial institution. You have the power to engage your communities, to help bring more people into the work. If you can, donate your time, skills, money or passion to youth- and Indigenous-led organizations.

Zoe could only save the boat because she was mentored, prepared and supported by generations upon generations before her. Let's create the same intergenerational bonds as we collectively work towards climate justice. Together, we can, and must, save the boat.

Endnotes

1. R. J. Brulle, 'Institutionalizing delay: foundation funding and the creation of US climate change counter-movement organizations', *Climatic Change*, 2014, 122, 681–94, https://doi.org/10.1007/s10584-013-1018-7; R. J. Brulle, 'The climate lobby: a sectoral analysis of lobbying spending on climate change in the USA, 2000 to 2016', *Climatic Change*, 2018, 149, 289–303, https://doi.org/10.1007/s10584-018-2241-z

2. G. Supran and N. Oreskes, 'Assessing ExxonMobil's climate change communications (1977–2014)', *Environmental Research Letters*, 2017, 12, 084019, https://doi.org/10.1088/1748-9326/aa815f

3. A. Lieberman and S. Rust, 'Big Oil braced for global warming while it fought regulations', *The Los Angeles Times*, 31 December 2015, http://graphics.latimes.com/oil-operations/

4. Brulle, 'The climate lobby', 2018; A. Sharp, 'Oil and gas majors have spent $1 billion undermining climate action since 2015, report says', *National Observer*, 21 March 2019, https://www.nationalobserver.com/2019/03/21/news/oil-and-gas-majors-have-spent-1-billion-undermining-climate-action-2015-report-says

5. See also 'Mother Earth' by Deborah McGregor in this volume.

Index

F

fertilizer 185, 197–199, 218–219, 226–227

fire. *See* bushfire; *See* wildfire

First Nations. *See* Indigenous peoples

fishery 5, 10, 16, 178, 179, 180, 181, 182, 183, 184. *See also* overfishing

Fleming, R. H. 204

flooding 44, 64–65, 143, 154, 164–165

Food and Agricultural Organization (FAO) 178–179

forest 3, 5, 15–16, 55–56, 67–73, 88, 90, 137–138, 161, 178, 196–197, 199

forest management 72

forest mortality 69

Forest Principles 153

fossil fuel 5, 28–30, 39–40, 46–48, 61, 70, 81, 90–92, 142, 144, 156–157, 162, 174, 177–178, 196, 205, 214–215, 248–249

Frank R. Lautenberg Chemical Safety for the 21st Century Act 189

fresh water 5, 101, 199, 221, 223–227

Fridays for Future 174

Friends of the Earth 33

Fukushima Daiichi nuclear disaster 60

G

G-7 Summit 37

Gandhi, Indira 35

General Agreement on Tariffs and Trade (GATT) 37–38

genetically modified (GM) crop 200

geoengineering 5, 19, 49, 91, 93, 95–99

glacier. *See* cryosphere: glacier

Global Assessment Report on Biodiversity and Ecosystem Services 151

global corporate rule 34, 38, 40

Global South. *See* North–South divide

global warming 16, 23–25, 27–29, 39, 91, 93–95, 107, 145, 151, 156, 162–167

Gore, Al 233–234
   An Inconvenient Truth 234

Graedel, Tom 192

Grandjean, Philippe 187

Great Seattle Fire, the 147

Great Smog, the 82–83, 85–86

Great Stink, the 48

greenhouse gas 3, 24–30, 39, 62–65, 70–74, 85, 91, 93, 96, 98–99, 105–107, 144, 151, 154–156, 172–174, 199, 214, 218

Greenland 24, 29, 102–106, 142–143, 208

Green New Deal 145–146

green politics 39

Green Revolution, the 185, 197–199

Green Science Policy Institute 190

greenwashing 55

Grotius, Hugo 204

Gutenberg, Beno 142

H

Haber-Bosch process 197

Hansen, James 231

Hardin, Garrett 78–79, 181
    'The tragedy of the commons' 78, 181

'Harvard Six Cities' study 89

heat wave 19, 68, 161–163, 167

high Earth orbit (HEO) 241–242

HMS *Challenger* 204

Holocene, the 104

hurricane 64–65, 148, 233–235

Hurricane Harvey 233

Hurricane Katrina 145, 147, 234

Hurricane Maria 64

Hurricane Rita 234

Hurricane Sandy 144

hydropower 56, 221, 225

I

ice sheet. *See* cryosphere: ice sheet

incineration 216–217

Indigenous knowledge systems (IKS) 5, 133–134, 136–137

Indigenous peoples 4–6, 133–138, 235–236, 248–250

Indigenous Peoples Earth Charter. *See* Kari-Oca Declaration

individual transferable quota (ITQ) 182

Industrial Revolution, the 53, 177, 196

Intergovernmental Panel on Climate Change (IPCC) 3, 151, 153, 173, 207, 232
    Second Assessment Report 3

International Decade of Ocean Exploration 206

International Geophysical Year 87, 205, 241

International Space Station (ISS) 241

International Union for the Conservation of Nature (IUCN) 134

irrigation 56, 198–200, 221, 226

J

Johnson, Lyndon B. 95, 205

Johnson, M. W. 204

Joint Global Ocean Flux Study 206

Journal of Environmental Economics and Management 78

Author Biographies

Jon Abbatt is a Professor of Chemistry at the University of Toronto. He is an atmospheric chemist, working at the interface of atmospheric science and chemistry both in the laboratory and the field. His focus is on studies of the fundamental chemistry underlying environmental phenomena, including atmospheric ozone, urban haze and the roles of atmospheric aerosol particles in cloud formation and climate. Over the past few years, his attention has been largely on the chemistry of Arctic aerosols, wildfire emissions and the indoor environment. He is a fellow of the Royal Society of Canada and the American Geophysical Union.

Sally N. Aitken is a Professor of Forest and Conservation Sciences and Associate Dean Research and Innovation in the Faculty of Forestry at the University of British Columbia. Her research uses genomic tools, experimental plantings and climate models to understand the relationships between tree populations and climate, and determine the capacity for forests to adapt to new conditions. She advises governments on the management and conservation of forests in a changing climate, and co-authored the widely used textbook Conservation and the Genetics of Populations. She is an elected Fellow of the Royal Society of Canada.

David Archer is a Professor of Geophysical Sciences at the University of Chicago, and a popular science writer and educator. His research, teaching and writing focuses on the past, present and future of the global carbon cycle and the climate impacts of changing carbon dioxide and methane concentrations. His books include *The Climate Crisis: An Introductory Guide to Climate Change* (2009) and *Global Warming: Understanding the Forecast* (2006). He is a regular contributor to the RealClimate blog (a commentary site on climatology available at http://www.realclimate.org/), and has created several open-access online classes on climate change, which are available through Coursera (https://www.coursera.org/)

Edward Burtynsky is regarded as one of the world's most accomplished contemporary photographers. His works are included in the collections of over sixty major museums around the world. Burtynsky's distinctions include the inaugural TED Prize in 2005; the Governor General's Awards in Visual and Media Arts in 2016; the Outreach Award at the Rencontres d'Arles (2004); the Roloff Beny Book Award (2004); and the 2018 Photo London Master of Photography Award. Most recently, he has been named as the recipient of the 2019 Achievement in Documentary Award from the Lucie Foundation. He currently holds eight honorary doctorate degrees.

Candis Callison is an Associate Professor at the University of British Columbia in the School of Journalism, Writing and Media Studies and the Institute for Critical Indigenous Studies. She is the author of *How Climate Change Comes to Matter: The Communal Life of Facts* (2014) and the co-author, with Mary Lynn Young, of *Reckoning: Journalism's Limits and Possibilities* (2020). She belongs to the Tahltan Nation, located in what is now northwestern British Columbia, and is a regular contributor to the podcast, Media Indigena. She is also a member of the American Academy of Arts and Sciences and a Pierre Elliot Trudeau Foundation Fellow.

Zoe Craig-Sparrow is a member of the Musqueam Indian Band and was born and raised on the reserve in Vancouver, BC. She is a graduate of Political Science at the University of British Columbia, and also studied at the Paris Institute of Political Studies (Sciences Po) in France and Lester B. Pearson United World College of the Pacific on Vancouver Island. Craig-Sparrow is passionate about the rights of girls and women, and about the environment, and takes a particular interest in how these issues relate to Indigenous communities. In 2012, she traveled to the United Nations at the age of fifteen to present a submission to the UN Committee on the Rights of the Child. Craig-Sparrow is currently pursuing a Master's in Human Rights at the University of London and is the Co-Director of Justice for Girls (http://www.justiceforgirls.org/).

Gretchen C. Daily is the Bing Professor of Environmental Science in the Department of Biology, Director of the Center for Conservation Biology and senior fellow at the Woods Institute for the Environment, at Stanford University. She is co-founder and faculty director of the Stanford Natural Capital Project, a global partnership working to integrate the value nature provides to society into finance, development and conservation decisions. Her research is focused on understanding the dynamics of change in the biosphere, their implications for human well-being and the deep societal transformations needed to secure people and nature. She is an elected fellow of the US National Academy of Sciences, the American Academy of Arts and Sciences and the American Philosophical Society.

Julian Dowdeswell is Director of the Scott Polar Research Institute and Professor of Physical Geography in the University of Cambridge, UK. He is a glaciologist, working on the form and flow of glaciers and ice sheets and their responses to climate change, and the links between former ice sheets and the marine geological record, using satellite, airborne and shipborne geophysical tools. Over the past four decades he has worked many times in Antarctica and the Arctic. He was awarded the Polar Medal by HM The Queen for 'outstanding contributions to glacier geophysics' (1994) and has also received the Founder's Gold Medal of the Royal Geographical Society (2008). As well as many scientific papers, he has co-written the popular science books *The Continent of Antarctica* (2018) and *Islands of the Arctic* (2002).

Don Fullerton is Gutgsell Professor in the Finance Department at the University of Illinois at Urbana-Champaign. His research analyzes distributional and efficiency effects of environmental and tax policies. He is the former Editor of the *Journal of the Association of Environmental and Resource Economists* and former Director of the Environmental and Energy Economics research program of the National Bureau of Economic Research. After taking a BA from Cornell University and a PhD in Economics from the University of California, Berkeley, he taught at Princeton University, the University of Virginia, Carnegie Mellon University and the University of Texas, before joining the University of Illinois in 2008. From 1985 to 1987, he served in the US Treasury Department as Deputy Assistant Secretary for Tax Analysis.

Roland Geyer is a Professor of Industrial Ecology at the Bren School of Environmental Science Management, University of California, Santa Barbara. Prior to joining the Bren School, he held research positions in Germany, France and the UK. Since 2000 he has worked with a wide range of governmental organisations, trade associations and companies on environmental sustainability issues. Geyer has won multiple awards for his work, such as the Royal Statistical Society's International Statistic of the Year (2018), and has been featured widely in the media, on programs such as CBS' *60 Minutes*, *CBS News Sunday Morning* and *PBS NewsHour*. He has a graduate degree in physics and a PhD in engineering. Learn more about Geyer and his work on his website (www.rolandgeyer.com).

Alice Gorman is an Associate Professor in the College of Humanities, Arts and Social Sciences at Flinders University in Adelaide, Australia. Her research focuses on the archaeological record of humans in outer space, including orbital debris, planetary landing sites and deep space probes. She is a Director on the Board of the JustSpace Alliance, a member of the Advisory Council of the Space Industry Association of Australia and a Senior Member of the American Institute of Aeronautics and Astronautics. Her 2019 book *Dr Space Junk vs the Universe: Archaeology and the Future* received the John Mulvaney Book Award and the NIB Literary Award People's Choice.

John Harte is a Professor of the Graduate School at the University of California, Berkeley. His research focuses on climate change, biodiversity and maintaining ecosystem services for humanity. He has authored over 240 scientific publications, including eight books, one of which, *Consider a Spherical Cow: A Course in Environmental Problem Solving* (1985) is a widely used textbook on environmental modeling. Along with Robert Socolow, he co-edited the 1970 book, *Patient Earth*. He is an elected Fellow of the Ecological Society of America, the American Physical Society and the American Association for the Advancement of Science, and a recipient of a Guggenheim Fellowship, the Leo Szilard prize from the American Physical Society, and a 2006 George Polk award in investigative journalism.

Janet G. Hering is a Professor of Biogeochemistry at ETH Zurich and of Environmental Chemistry at the École Polytechnique Fédérale de Lausanne, and Director of the Swiss

Federal Institute of Aquatic Science and Technology (EAWAG). Her research examines the biogeochemical cycling of trace elements in water, and knowledge exchange at the interface of science with policy and practice. She has worked closely with various government agencies to address issues of arsenic contamination in drinking water, and has served on the Advisory Board of the US Environmental Protection Agency. She is the co-author, with François Morel, of the highly influential textbook, *Principles and Applications of Aquatic Chemistry* (1993), and is an elected member of the US National Academy of Engineering.

Sheila Jasanoff is a Professor of Science and Technology Studies at the John F. Kennedy School of Government at Harvard University, and the founding Director of the Program on Science, Technology and Society. Her work focuses on the interactions between science and the state in contemporary democratic societies, with implications for comparative politics, political theory, law and sociology. She has conducted comparative research in the United States, the United Kingdom, Germany, the European Union and India, as well as on emerging global regimes in areas such as climate and biotechnology. She is the author of many influential books, including *The Ethics of Invention: Technology and the Human Future* (2016) and *The Fifth Branch: Science Advisers as Policymakers* (1990).

David M. Karl is a Professor of Oceanography at the University of Hawai'i, and Director of the Center for Microbial Oceanography: Research and Education (C-MORE). His research explores global-scale changes in ocean properties, and, in 1988, he established the Hawaii Ocean Time-series (HOT) program to conduct sustained oceanographic measurements in the subtropical Pacific waters near Hawai'i. Since 1973, he has spent more than 1,000 days conducting research at sea, including twenty-three trips to Antarctica. He is a Fellow of the American Geophysical Union and the American Academy of Microbiology, and a member of the US National Academy of Sciences.

Robert E. Kopp is the Director of the Institute of Earth, Ocean and Atmospheric Sciences at Rutgers University, and a Professor in the Department of Earth and Planetary Sciences. His research focuses on past and future sea-level change, on the interactions between physical climate change and the economy, and on the use of climate risk information in decision making. He is an author of *Economic Risks of Climate Change: An American Prospectus* (2015), the Fourth National Climate Assessment (2017), and the Intergovernmental Panel on Climate Change's Sixth Assessment Report (forthcoming, 2021). He is a Fellow of the American Geophysical Union and a recipient of its James B. Macelwane Medal (2017).

Rosemary Lyster is a Professor of Climate and Environmental Law at The University of Sydney Law School, and a Fellow of the Australian Academy of Law. Her research focuses on Climate Law, Climate Justice and Disaster Law. She has published many articles and seven books, including most recently *Research Handbook on Climate Disaster Law: Barriers and Opportunities* (edited with Robert Verchick, 2018) and *Climate Justice and Disaster Law* (2015). In 2018, Rosemary was identified by the *Australian Financial Review* as one of Australia's '100 Women of Influence'. In 2013, she was appointed a Herbert Smith Freehills Visiting Professor at the Faculty of Law, University of Cambridge, and was a Visiting Scholar at Trinity College, Cambridge in 2009 and in 2014.

Douglas G. MacMartin is a Senior Research Fellow in the Sibley School of Mechanical and Aerospace Engineering at Cornell University. His research focuses on climate engineering, with the aim of helping to develop the knowledge base necessary to support informed future societal decisions in this challenging and controversial field. He has published extensively on the subject, and given many public and academic presentations, including briefings to the UN Environment Program and testimony to the US Congress. He is a member of a US National Academies panel that will make recommendations on climate engineering research and governance.

Elizabeth May is the Green Party of Canada's first elected Member of Parliament. Prior to her career in politics, she practiced law with the Public Interest Advocacy Centre, served as the Senior Policy Advisor to the federal Canadian Minister of the Environment (1986–1988), and as Executive Director of the Sierra Club of Canada (1989–2006). Elizabeth is the author of eight books, including, most recently, *Who We Are: Reflections on My Life and on Canada* (2014). In November 2019, May stepped down as leader of the Green Party of Canada after serving for thirteen years, and continues as Parliamentary Leader for Canada's first federal

Green Party Caucus. She was inducted as an Officer of the Order of Canada in recognition of her decades of leadership in the Canadian environmental movement.

Deborah McGregor is an Associate Professor and Canada Research Chair in Indigenous Environmental Justice at York University in Toronto. Her research focuses on Indigenous knowledge systems and their applications in diverse contexts, including water and environmental governance, environmental justice, forest policy and management and sustainable development. She has delivered numerous public and academic presentations on the topics of Indigenous knowledge systems, governance and sustainability, and she co-edited the book *Indigenous Peoples and Autonomy: Insights for a Global Age* (2010). McGregor is Anishinaabe from Whitefish River First Nation, Birch Island, Ontario.

Neville Nicholls is an Emeritus Professor in the School of Earth, Atmosphere and Environment at Monash University. He spent thirty-five years as a research scientist in the Australian Bureau of Meteorology before moving to Monash in 2006. He has studied the nature, causes, impacts and predictability of weather and climate extremes, especially droughts, heatwaves, tropical cyclones, floods and cold fronts, both globally and across the Australian region. Nicholls is a Fellow of the Australian Academy of Science.

Elias Grove Nielsen is a Danish entrepreneur who has co-founded many companies, including Biomega.com, Seed.dk and E-invasion.com. Innovative bicycles designed by Biomega.com have been exhibited in Centre Pompidou (Paris), Neues Museum (Nürnberg) and the San Francisco MoMa (Permanent Collection). He recently sold his businesses and now writes full-time.

Grace Nosek is the Founder and Student Director of the University of British Columbia Climate Hub. She is a PhD candidate at the Peter A. Allard School of Law, studying how to use law to protect climate change science from manufactured doubt. She is a Pierre Elliott Trudeau Foundation scholar and a Killam doctoral scholar, and a past Canada-US Fulbright recipient. She holds a BA from Rice University, a JD from Harvard Law School, and an LLM from the University of British Columbia. Drawing from her research, Nosek creates hopeful climate narratives, including the *Ava of the Gaia* trilogy (2011, 2014, 2018), the short film *Climate Comeback* (2019) and the podcast series *Planet Potluck* (2018–present).

Daniel Pauly is a Professor of Fisheries at the University of British Columbia, Vancouver Canada, where he directs the Sea Around Us project, which is devoted to studying, documenting and mitigating the impact of fisheries on the world's marine ecosystems. The concepts, methods and software his team has developed have been used in over 1000 widely-cited publications, and have led to his receiving numerous scientific awards.

Navin Ramankutty is a Professor and Canada Research Chair in Global Environmental Change and Food Security at the University of British Columbia. His research uses global data and models to explore solutions to improve global food system sustainability. He is a Scientific Steering Committee member of the Global Land Programme. He was a lead author of the Millennium Ecosystem Assessment (2005), and a contributing author of the Fourth Assessment Report of the Intergovernmental Panel on Climate Change (2007), and is currently contributing to the Intergovernmental Science-Policy Platform on Biodiversity and Ecosystem Services. He is a Leopold Leadership Fellow.

Katharine L. Ricke is an Assistant Professor at the University of California, San Diego, jointly appointed between the School of Global Policy and Strategy, and the Scripps Institution of Oceanography. She is an interdisciplinary climate scientist who applies quantitative modeling and large data analysis methods to social and physical systems. Her research focuses on how uncertainty and heterogeneity in projected climate impacts and solutions, influence strategic incentives in climate policy problems. She is a 2019 Andrew Carnegie Fellow, and a member of the US National Academy committee that will make recommendations on climate engineering research and governance.

Tapio Schneider is a Professor of Environmental Science and Engineering at the California Institute of Technology (Caltech), and a Senior Research Scientist at NASA's Jet Propulsion Laboratory. His research examines the impacts of climate change on extreme rainfall, the role of cloud cover in the climate system, and the factors creating winds and weather on planetary bodies such as Jupiter and Titan. He currently leads the Climate Modeling Alliance (https://clima.caltech.edu/), whose mission is to build the first Earth system model that automatically learns from diverse data sources to produce accurate climate predictions. He was named one of the '20 Best Brains Under 40' by *Discover* in 2008, was a David and Lucile Packard Fellow and Alfred P. Sloan Foundation Research Fellow, and a recipient of the James R. Holton Award of the American Geophysical Union (2004).

Jeffrey R. Smith is a PhD candidate in the Department of Biological Sciences at Stanford University and the Stanford Center for Conservation Biology. His research focuses on the roles of climate and human land use in driving patterns of biodiversity and ecological interactions, with an emphasis on insect diversity. He is a National Science Foundation Graduate Research Fellow and a Beijer Young Scholar at the Beijer Institute for Ecological Economics. He received a Master in Environmental Science (MESc) from the Yale School of Forestry and Environmental Studies and a BSc from the University of Delaware.

Robert Socolow is Professor Emeritus in the Department of Mechanical and Aerospace Engineering at Princeton University. He was trained in theoretical high-energy physics, and began his career as an Assistant Professor of physics at Yale Univeristy, before joining the Princeton faculty in 1971 with the assignment of inventing interdisciplinary environmental research. That same year, he wrote and edited, with John Harte, *Patient Earth*, an early textbook in environmental studies. He has since published many highly influential papers, including the 2004 *Science* paper, with Steve Pacala, 'Stabilization wedges: Solving the climate problem for the next 50 years with current Technologies'. Rob's work has also introduced 'one billion high emitters', 'committed emissions' and 'destiny studies', as further conceptual frameworks for climate change policy. He is a member of the American Academy of Arts and Sciences, a fellow of the American Physical Society and a fellow of the American Association for the Advancement of Science. Between 1992 and 2002, he was the editor of *Annual Review of Energy and the Environment*.

U. Rashid Sumaila is a Professor and Director of the Fisheries Economics Research Unit at the Institute for the Oceans and Fisheries at the University of British Columbia (UBC), and Director of the OceanCanada Partnership. He is also appointed in the UBC School of Public Policy and Global Affairs. His research focuses on bioeconomics, marine ecosystem valuation and the analysis of global issues such as fisheries subsidies, illegal fishing, climate change and oil spills. Sumaila has published widely, and is on the Editorial Boards of several journals, including *Science Advances*, *Scientific Reports* and *Environmental and Resource Economics*. He was the recipient of the 2017 Volvo Environment Prize and other prestigious awards, and is a Fellow of the Royal Society of Canada. Sumaila has given addresses at the UN Rio+20 meeting, the World Trade Organization, the White House, the Canadian Parliament, the African Union, St James's Palace and the British House of Lords.

Elsie Sunderland is the Gordon McKay Professor of Environmental Chemistry at Harvard University. She holds faculty appointments in the Harvard John A. Paulson School of Engineering and Applied Sciences, the Harvard T.H. Chan School of Public Health, and the Department of Earth and Planetary Sciences. She is a faculty associate in the Harvard University Center for the Environment (HUCE) and the Harvard Center for Risk Analysis (HCRA). Her research focuses on how releases of persistent environmental contaminants are transformed by the physical and biological processes in the environment, and how this affects human exposures and risk of adverse health outcomes. Prior to joining the faculty at Harvard, she spent five years working to develop science-based environmental policy at the US Environmental Protection Agency, including regulatory impact assessments and guidance on the use of environmental models to inform regulatory decisions.

Philippe Tortell is a Professor at the University of British Columbia (UBC), and the Head of the Department of Earth, Ocean and Atmospheric Science. He is a sea-going oceanographer, with more than two decades of experience documenting the effects climate change on marine systems around the world. His research examines the impacts of changing ocean conditions on marine biological productivity and the biogeochemical cycling of climate-active gases. He is member of the College of the Royal Society of Canada, a Von Humboldt Research Fellow and past Director of UBC's Institute for Advanced Studies. He has previously edited two inter-disciplinary books — *Reflections of Canada* (2017) and *Memory* (2018).

Charlotte C. Wagner is a PhD researcher at the John A. Paulson School of Engineering and Applied Sciences at Harvard University, where she studies the fate and transport of globally distributed man-made pollutants in the environment. Her research combines environmental engineering, biogeochemistry and environmental health to better understand the long-term human and ecological exposure risks of environmental pollution. Prior to starting her doctorate, Wagner worked for several years as a scientific writer and editor, raising awareness of health risks arising from chemical contaminants. She received a BA in Political Science and Environmental Policy from Maastricht University, and a Master of Science degree in Environmental Health from the Cyprus International Institute for Environmental and Public Health.

Hannah Wittman is a Professor in the Institute of Resources, Environment and Sustainability, and Academic Director of the Centre for Sustainable Food Systems at the University of British Columbia. She conducts community-based and participatory action research related to food sovereignty, agrarian reform, agroecology and health equity in Canada and Latin America. She is also co-Specialty Chief editor of the Social Movements, Institutions and Governance section of *Frontiers in Sustainable Food Systems*. Her edited books include *Environment and Citizenship in Latin America: Natures, Subjects and Struggles* (2012), *Food Sovereignty in Canada: Creating Just and Sustainable Food Systems* (2011) and *Food Sovereignty: Reconnecting Food, Nature and Community* (2010).

Elizabeth J. Wilson is a Professor of Environmental Studies at Dartmouth College and the founding Director of the Arthur L. Irving Institute for Energy and Society at Dartmouth. Her research focuses on energy system transitions, including interactions between technologies, decision making, policies and institutions. She works with practitioners in government, civil society and the private sector to shape the next generation of energy system transitions. She was awarded a Carnegie Fellowship and was selected as a Leopold Leadership Fellow. She is also a board member at the Vermont Energy Investment Corporation.

The Publishing Team

Alessandra Tosi was the managing editor for this book.

Adele Kreager performed the copy-editing, proofreading and the indexing.

Anna Gatti designed the cover using InDesign. The image featured on the cover is *Earthrise* (24 December 1968), a photo taken by Apollo 8 crewmember Bill Anders (source: https://commons.wikimedia.org/wiki/File:NASA_Earthrise_AS08-14-2383_ Apollo_8_1968-12-24.jpg). The cover was produced in InDesign using Myriad (titles) and Calibri (text body) fonts.

Luca Baffa typeset the book in InDesign. The text is set in Libre Baskerville. Luca created all of the editions — paperback, hardback, EPUB, MOBI, PDF, HTML, and XML. The conversion was performed with open source software freely available on our GitHub page (https://github.com/OpenBookPublishers).

For information on the environmental impact of Open Book Publishers, see https:// blogs.openbookpublishers.com/environmental-impact

This book need not end here...

Share

All our books — including the one you have just read — are free to access online so that students, researchers and members of the public who can't afford a printed edition will have access to the same ideas. This title will be accessed online by hundreds of readers each month across the globe: why not share the link so that someone you know is one of them?

This book and additional content is available at: https://doi.org/10.11647/OBP.0193

Customise

Personalise your copy of this book or design new books using OBP and third-party material. Take chapters or whole books from our published list and make a special edition, a new anthology or an illuminating coursepack. Each customised edition will be produced as a paperback and a downloadable PDF.

Find out more at: https://www.openbookpublishers.com/section/59/1

You may also be interested in:

What Works in Conservation 2020

William J. Sutherland, Lynn V. Dicks, Silviu O. Petrovan and Rebecca K. Smith (eds.)

https://doi.org/10.11647/OBP.0191

Living Earth Community
Multiple Ways of Being and Knowing

Sam Mickey, Mary Evelyn Tucker, and John Grim (eds.)

https://doi.org/10.11647/OBP.0186

Forests and Food
Addressing Hunger and Nutrition Across Sustainable Landscapes

Bhaskar Vira, Christoph Wildburger and Stephanie Mansourian (eds.)

https://doi.org/10.11647/OBP.0085

CPSIA information can be obtained
at www.ICGtesting.com
Printed in the USA
LVHW071148290420
654553LV00015B/390

9 781783 748457